Viktor K. Jirsa · J. A. Scott Kelso (Eds.)

Coordination Dynamics:
Issues and Trends

With 90 Figures

Springer

Series Editor:

Prof. J. A. Scott Kelso
Florida Atlantic University
Center for Complex Systems
Glades Road 777
33431-0991 Boca Raton FL / USA

Volume Editors:

Dr. Viktor K. Jirsa
Prof. J. A. Scott Kelso
Florida Atlantic University
Center for Complex Systems
Glades Road 777
33431-0991 Boca Raton FL / USA
jirsa@walt.ccs.fau.edu
kelso@ccs.fau.edu

ISBN 3-540-20323-0 Springer-Verlag Berlin Heidelberg New York

Cataloging-in-Publication Data applied for
Bibliographic information published by Die Deutsche Bibliothek.
Die Deutsche Bibliothek lists this publication in the Deutsche Nationalbibliografie;
detailed bibliographic data is available in the Internet at <http://dnb.ddb.de>.

This work is subject to copyright. All rights are reserved, whether the whole or part of the material is concerned, specifically the rights of translation, reprinting, reuse of illustrations, recitation, broadcasting, reproduction on microfilm or in other ways, and storage in data banks. Duplication of this publication or parts thereof is permitted only under the provisions of the German Copyright Law of September 9, 1965, in its current version, and permission for use must always be obtained from Springer-Verlag. Violations are liable for prosecution under German Copyright Law.

Springer-Verlag is a part of Springer Science+Business Media

springeronline.com

© Springer-Verlag Berlin Heidelberg 2004
Printed in Germany

The use of general descriptive names, registered names, trademarks, etc. in this publication does not imply, even in the absence of a specific statement, that such names are exempt from the relevant protective laws and regulations and therefore free for general use.

Typesetting: Dataconversion by editors
Cover-design: E. Kirchner, Heidelberg
Printed on acid-free paper   62 / 3020 hu - 5 4 3 2 1 0

Understanding Complex Systems

Series Editor: J. A. Scott Kelso

Springer
Berlin
Heidelberg
New York
Hong Kong
London
Milan
Paris
Tokyo

Springer Complexity

Springer Complexity is a publication program "cutting across all traditional disciplines of sciences as well as engineering, economics, medicine, psychology and computer sciences, which is aimed at researchers, students and practitioners working in the field of complex systems". Complex Systems are systems that comprise many interacting parts with the ability to generate a new quality of macroscopic collective behavior through self-organization, e.g. the spontaneous formation of temporal, spatial or functional structures. This recognition, that the collective behavior of the whole system cannot be simply inferred from the understanding of the behavior of the individual components, has led to various new concepts and sophisticated tools of complexity. The main concepts and tools - with sometimes overlapping contents and methodologies are the theories of self-organization, complex systems, synergetics, dynamical systems, turbulence, catastrophes, instabilities, nonlinearity, stochastic processes, chaos, neural networks, cellular automata, adaptive systems, or genetic algorithms.

The topics treated within Springer Complexity are as diverse as lasers or fluids in physics, cutting phenomena of workpieces or electric circuits with feedback in engineering, growth of crystals or pattern formation in chemistry, morphogenesis in biology, brain functions in neurology, behavior of stock exchange rates in economics, or the formation of public opinion in sociology. All these seemingly quite different kinds of structure formation have a number of important features and underlying structures in common. These deep structural similarities can be exploited to transfer analytical methods and understanding from one field to another. Therefore the Springer Complexity program seeks to foster cross-fertilization between the disciplines and a dialogue between theoreticians and experimentalists for a deeper understanding of the general structure and behavior of complex systems.

The program consists of individual books, books series such as "Springer Series in Synergetics," "Institute of Nonlinear Science," "Physics of Neural Networks," and "Understanding Complex Systems," as well as various journals.

Preface

The Problem of Coordination

Coordination represents one of the most striking, most taken for granted, yet least understood features of all living things. Imagine a living system whose component parts and processes, on any level of description one chooses to examine, did not interact with each other or with their surrounds. Such a "cell," "organ," "organism" or "factory" would possess neither structure nor function. Howard Pattee, one of the deepest thinkers of our time—a true "conceptual biologist"—goes so far as to say that he does " ...*not see any way to avoid the problem of coordination and still understand the physical basis of life*". How are complex living things coordinated in space and time? Coordination is not just matter in motion, rather, coordination is a functional spatiotemporal order. This order can be seen almost everywhere we look, whether it be in the regulatory interactions among genes that affect how an organism develops and how some diseases like cancer occur, the coordinated responses of organisms to constantly varying environmental stimuli, the coordination among nerve cells of the brain that underlie our ability to think, act and remember, the miraculous coordination that belies a child's first word, the fingers of the pianist playing a concerto, people working together to achieve a common goal, ...Look around you. Everything we do is coordinated! Listen, for example, to the philosopher, Patricia Churchland, writing recently in the journal *Science* on "Self-Representation in Nervous Systems":

*"The most fundamental of the self-representational capacities probably arose as evolution stumbled on solutions for **coordinating** inner body signals to generate survival appropriate inner regulation. The basic **coordination** problems for all animals derive from the problem of what to do next. Pain signals should be **coordinated** with withdrawal, not with approach. Thirst signals should be **coordinated** with water seeking, not with fleeing unless a present threat takes higher priority. Homeostatic functions and the ability to switch between the different internal configuration (sic) for fight and flight from that needed for rest and digest require **coordinated** control of heart, lungs, viscera, liver, and adrenal medulla. Body-state signals have to be integrated, options evaluated, and choices made, since the organism needs to act as a coherent whole, not a group of independent systems with competing interests"* [i]

It is pretty clear in this quotation that a lot of science is tied up in the word coordination and its closely related kin—binding, recruiting, grouping, orchestrating, communicating, integrating, selecting, switching and so forth. But how do we understand these words? Are there rules or laws of coordination and if so what do they look like?

What is Coordination Dynamics?

Coordination Dynamics—the science of coordination-- describes, explains and predicts how patterns of coordination form, adapt, persist, and change in natural systems. Through an intimate, two-way relationship between theory and experiment, Coordination Dynamics seeks to identify the laws, principles and mechanisms underlying coordinated behavior in different kinds of system at different levels of description. It explicitly addresses coordination within and between levels of description and organization. Coordination Dynamics thus aims to characterize the nature of the coupling *within* a part of a system (e.g., the firing of cells in the heart or neurons in a part of the brain), *between* different parts of a system (e.g., parts of the brain, parts of the body, members of an audience) and *between* different kinds of systems (e.g., stimuli and responses, organisms and environments, perception and action, etc.). Ultimately Coordination Dynamics is concerned with how things come together in space and time, and how they split apart. *It is about discovering all kinds of dynamic patterns in the behavior of living things and expressing how these patterns form, adapt and change according to internal and external needs in terms of dynamical laws or rules*

What are the main ideas of Coordination Dynamics?

The first main idea is that basic patterns of coordination can—under certain conditions—arise spontaneously in a *self-organized* fashion. When people see patterns in the world or in themselves, they tend to think immediately that there must be some kind of centralized agent responsible for creating and orchestrating these patterns. It's akin to the mythological explanations of ancient times. Thunder literally meant the Norse God Thor's roar. This tendency to attribute events in the world to some *deus ex machina* is a difficult one to overcome (genes 'for' violence comes to mind). In self-organizing systems there is no homunculus-like agent located inside or instructions from the outside ordering the elements, telling them what to do and when to do it. Rather the system literally organizes itself. There is organization without an organizer, coordination without a coordinator, switching without switches, etc.

The second main idea is that the patterns that arise spontaneously due to self-organizing processes can be captured by *coordination or collective variables* that evolve in time. The patterns that emerge, persist and change thus possess a *pattern dynamics* capable, as we shall see, of generating a rich repertoire of behaviors. This, we hypothesize, is one of Nature's complementary themes: the processes of self-organization cause the formation of coordinated patterns of behavior. These patterns are governed by nonlinear laws that we collectively refer to as Coordination Dynamics. Of course, the big scientific problem in every field of endeavor is to identify the pattern or coordination variables and their pattern or coordination dynamics.

The third main idea is that Coordination Dynamics deals with *informational quantities* that transcend the medium through which the parts communicate. Evidence shows that things may be coupled by mechanical forces, by light, by sound, by smell, by touch and by intention. In Coordination Dynamics, "binding" or coupling is mediated by information, not—or not only—by conventional forces. Such information may not only be of a material but also of a structural or topological nature. It may cause qualitative changes in the dynamics of the coordinating parts and new states to emerge. Hence, 'bound' coordinative states in Coordination Dynamics are informational, and information that changes bound states is 'meaningful' to the system.

The fourth main idea is that Coordination Dynamics offers an explanation for *the origin of meaningful information*. We are used to thinking about information in terms of binary digits (bits). If I have a coin in my hand and ask you whether it is a head or a tail, the answer requires one bit of information. Choosing between any two alternatives—no matter what they are about—also takes one bit of information. The context or the meaning of the message is irrelevant. A bit is a bit regardless of context. In Coordination Dynamics new information is created because the system operates in a special (metastable) régime in which the notion of a state must be extended to that of a transient state, or, even better, to the notion of (coexisting) *tendencies*! These tendencies are for the (context-dependent) parts to retain their autonomy or independence (a segregating, diverging tendency), and the simultaneous tendency for the parts to coordinate together (an integrating, converging tendency). In the metastable régime, the slightest nudge will put the system into a new coordinated or 'bound' state. In this way, the (essentially nonlinear) coordination dynamics creates new information that can be stabilized over time. The *stability of information over time* is guaranteed by the coupling between component parts and processes and constitutes a dynamic kind of (non-hereditary) memory.

A corollary of the fourth main idea concerns *the biological origin of agency and consciousness*. Self-organizing processes may well provide the foundation of all forms of coordination. However, we do not want to throw the baby out with the bathwater. Coordinated behavior often has a goal-directedness to it as well. Where do agency and directedness come from? A central hypothesis of Coordination Dynamics is that spontaneous self-organizing coordination tendencies give rise to agency; that the most fundamental kind of consciousness, the awareness of self, springs from the ground of spontaneous self-organized activity. Think, for example, of the elementary spontaneous movements we are born with—the pre-existing repertoire--such as making a fist, kicking, sucking, stretching, etc. Such activities happen before we make them happen. *Spontaneous (self-organized) coordination tendencies thus lie at the origins of conscious agency.* They are, in the words of the philosopher Maxine Sheets-Johnstone, "the mother of all cognition", presaging every conscious mind that ever said "I".

The fifth main idea is that *information once created and 'stored' can direct, guide and modify* the
coordination dynamics. Capability gives rise to possibility. A great benefit of knowing a system's spontaneous, self-organizing tendencies (its so-called intrinsic dynamics) is that this provides the key information about which coordination variables can be modified (for example by learning and development) and which cannot. We are used to thinking about information in terms of codes (the 'genetic code', for instance) or as language communicated between individuals. In Coordination Dynamics, meaningful information arises from the dynamical coupling between things (typically a product of co-evolution), and it is this information that can be used to guide and direct ongoing coordinative activity.

The sixth main idea concerns *what information does and how it does it*. In Coordination Dynamics, information plays a specific (and once again dual) role. On the one hand, information may *stabilize* coordination states under conditions in which they are unstable and susceptible to global change—such as undergoing a phase transition. On the other hand, information can *destabilize* coordination states in order to fit the needs of the organism or the current demands of the situation. Specific information may come from environmental or internal sources. For example, specific ('epigenetic') information appears responsible for turning on and off the expression of genes, thus affecting whether a cell becomes a kidney or a heart cell, and ultimately affecting what organs do. In short, the role of information is to both stabilize and destabilize the dynamics depending on context.

The seventh main idea is that Coordination Dynamics offers a *way to connect levels of organization*. The key idea is that no matter what the level of organization investigated, it is possible to identify laws or rules of coordination: equations of motion whose parameters alter the stability and change of coordination states over time. At the next level down, individual components and the dynamics of component producing processes can be identified. Coordination emerges from the nonlinear interactions among component processes under appropriate conditions. In this way, lawful coordination behavior at one level is 'constructed' from the couplings among components. We call this dual scientific strategy of reducing down and constructing up, 'constructive reductionism'.

The eighth main idea is that Coordination Dynamics offers *a vocabulary and a language for understanding coordination especially in the world of living things*. Just as Quantum Mechanics at the beginning of the last century transformed the language of the physical world by replacing mass particles that had definite position and momentum with probabilities and wave functions, so the vocabulary of nonlinear dynamical systems offers a way to describe the way things behave and evolve in time. As a mathematical toolbox, nonlinear dynamics becomes science when it is filled with content, namely the key variables and parameters involved in specific coordinative processes. We remark again how ubiquitous is coordinative activity, for example between genes and proteins, within and between different regions of the brain, between an organism and its environment. Every system is different but the intriguing possibility is that what we learn about the Coordination Dynamics of one may aid in understanding another.

In this volume we brought scientists together from all over the world who have defined and developed the field of Coordination Dynamics. The contributed articles are collapsed in five parts. Part I investigates the philosophical foundations of Coordination Dynamics. In Part II its cognitive contributions are discussed such as attention, intention and learning. Part III contains discussions of control mechanisms present in postural control. Part IV investigates the degree of perceptual versus motoric influences in Coordination Dynamics. Finally, general aspects of integration and segregation present in Coordination Dynamics are discussed in Part V.

The realization of this volume would have not been possible without the help of Drs Julien Lagarde and Olivier Oullier who did most of the proof-reading and indexing. Our thanks also go to Ajay Pillai who did most of the formatting. We are grateful to our sponsors for supporting the conference that brought the contributing scientists together and led to this book: National Institute of Mental Health, Florida Atlantic University Foundation, Florida Atlantic University Office of the Vice President for Research, and University MRI. Finally, we would like to thank Dr Tom Ditzinger of Springer-Verlag for his guidance and support throughout the production of this book.

Boca Raton, Florida.　　　　　　　　　　　　　　　　　　　　　　　J.A. Scott Kelso
August 11th, 2003　　　　　　　　　　　　　　　　　　　　　　　　Viktor K. Jirsa

Table of Contents

Part I: Philosophical Investigations of Coordination Dynamics: Perception and Action

    Impredicativity, Dynamics, and the Perception-Action Divide 1

Part II: Cognitive Contributions to Coordination Dynamics: Attention, Intention and Learning

    A Dynamical Approach to the Interplay of Attention and Bimanual Coordination 21
    Intention in Bimanual Coordination Performance and Learning 41
    Searching for (Dynamic) Principles of Learning 57

Part III: Coordination Dynamics of Posture: Control Mechanisms

    Using Visual Information in Functional Stabilization: Pole-Balancing Example 91
    Postural Coordination Dynamics in Standing Humans.............................103
    Noise Associated with the Process of Fusing Multisensory Information...123

Part IV: Perceptual and Motoric Influences on Coordination Dynamics

    Governing Coordination. Why do Muscles Matter?...................................141
    Guiding Movements without Redundancy Problems 155
    A Perceptual-Cognitive Approach to Bimanual Coordination 177

Part V: Integration and Segregation in Coordination Dynamics

    Complex Neural Dynamics...197
    Oscillations and Synchrony in Cognition...217
    Integration and Segregation of Perceptual and Motor Behavior 243

Author Index 261

Index 265

Part I: Philosophical Investigations of Coordination Dynamics: Perception and Action

Impredicativity, Dynamics, and the Perception-Action Divide

M. T. Turvey

Center for the Ecological Study of Perception and Action, University of Connecticut, Storrs CT, USA and Haskins Laboratories, New Haven CT, USA

In this brief and largely pictorial essay I address the divide between perception and action. I review theoretical perspectives on ways in which the divide might be crossed and on ways in which the divide might be dissolved. Some of the ways to either cross or dissolve are traditional, others are recent, and some, importantly for our present purposes, are fundamentally dynamical.

At the core of this essay is a stream of unconventional ideas. It flows from two sources: Robert Rosen's "The epistemology of complexity" and Gregor Schöner's and Scott Kelso's "Dynamic patterns of biological coordination." These were two papers presented 15 years ago at the conference on Dynamical Patterns in Complex Systems honoring Hermann Haken on his 60th birthday. They appeared in the published proceedings (Kelso, Mandell, & Shlesinger, 1988). An undercurrent in this stream of ideas is my ruminations on the aforementioned two papers. I was intrigued and tantalized by them. I was also perplexed and vexed by them. They were, for me, the most important papers at the conference but I doubted then, and I doubt now, that their respective theses were understood or understandable. My impression, 15 years delayed, is that the desired and invaluable comprehension is attainable only by considering the two papers jointly and only by paying homage to the nexus of ideas that are their historical backdrop. Ideally, this pictorial essay moves us a little way toward the latter goal.

1. Historical reasons for the perception-action divide

The perception-action divide is forced upon us by certain conceptions summarized in Figure 1. The top panel reviews Descartes's persistent influence, pictured in terms of his three substances (Descartes's "trialism") and the ghost-in-the-machine metaphor. The ghost's perception of the world outside his machine is based on a reasoned interpretation (3rd grade of sense) of the colors and sounds (2nd grade of sense) provided by a television set inside the machine's control room that receives inputs (1st grade of sense) from cameras and microphones attached to the machine's exterior. The ghost's actions are achieved via appendages and wheels attached to the machine's exterior. Whereas mechanical pushes and pulls constitute for Descartes the 1st grade of sense on the sensory side, hydraulic events

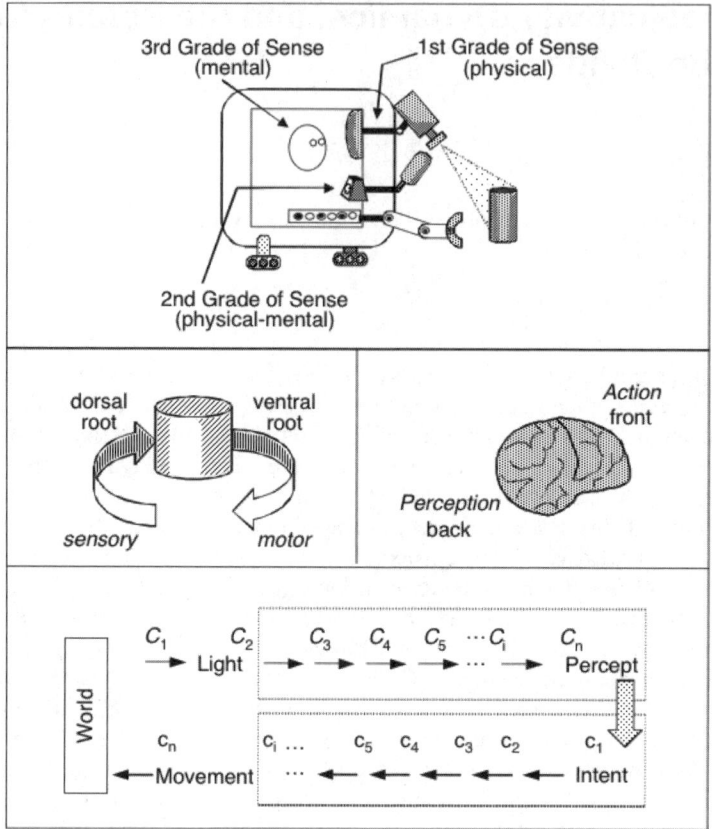

Figure 1. Historical reasons for the perception-action divide (see text)

constitute the 1st grade of sense on the motor side. By this metaphor, perception and action are mediated by the 3rd grade of sense.

Early 19th century research on the spinal cord culminated in the Bell-Magendie law depicted in the middle panel of Figure 1. Peripheral sensory and motor processes divide anatomically into the spine's dorsal and ventral roots, respectively. The law provided grounding for the basic Cartesian notion of *reflex arc*. The middle panel of Figure 1 also depicts a later triumph of the 19th century, following from observations of brain pathologies and consequences of electrical stimulation. The triumph was a rule of thumb: central perception and action processes divide at the central fissure of the cerebrum, action fore and perception aft.

The bottom panel of Figure 1 communicates a third persisting reason for the perception-action divide, one that accords with the core feature of Newton's physics: the equating of efficient cause with state transition. Thus, perception is envisaged as a linear causal chain (C_i) of transitional events arrayed spatially from world to percept and action is envisaged as a linear causal chain (c_i) of transitional events arrayed spatially from intent to world. (Dotted lines enclose causal brain states.)

2. Perspectives on how the perception-action divide is crossed

There are several theoretical perspectives on how to connect the logically separable domains of perception and action. Some are depicted in Figures 2 and 3. The left panel of Figure 2 depicts a very common perspective that might be simply labeled *translation*. It is a modern variant of the ghost-in-the-machine model seen in Figure 1. The physics of Descartes's 1st grades of sense on the sensory and motor sides and the states of the corresponding 2nd grades of sense are expressed in the terms of a Turing machine (see section 7). They retain, however, the Cartesian thesis that the sensory and motor processes are different in kind. In the Turing formulation the referents and governing syntax of sensory symbols are categorically distinct from those of motor symbols. The modern computational variant of the Cartesian 3rd grade of sense translates the sensory language into the motor language.

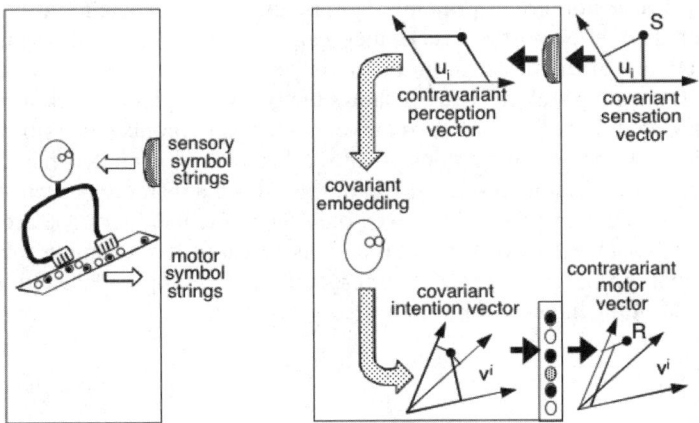

Figure 2. Two perspectives on how the perception-action divide is crossed when perception and action are taken to be incommensurable (see text)

A less common perspective is shown in the right panel of Figure 2. It interprets perception and action in terms of tensor transformations of coordinate systems (Pellionisz, 1985). The coordinate systems of perception and action are acknowledged as typically non-orthogonal (that is, not rectangular). At the top right of the Figure the environmental position of a distal object S is represented by independent sensory coordinates, u_i. Moving counterclockwise, the percept of S is represented in a unitary manner by interdependent coordinates, u^i. A metric tensor transforms the covariant vector representation of S into a contravariant vector representation.

Motor coordinate systems are presumed to have more dimensions than sensory coordinate systems. Covariant embedding is needed, therefore, to transform the percept vector into the vector representing the intention R with independent components (commands or, picturesquely, control buttons), v_i. The final transformation (via a metric tensor) is into the parallelogram-type motor vector R with components (muscle states) v^j. Let R be reaching. Then R can be a successful reach to S because the tensor transformations preserve the physical invariant of S's distance across all coordinate systems. In a word, the causal chains of perception and action are *transparent* to the physical facts of the environment in which they are embedded.

Whereas the panels of Figure 2 express the idea of perception and action as incommensurable, the panels of Figure 3 express the idea of perception and action as partially commensurable. The left panel portrays two hypothesized principles that bridge the divide: the common-coding principle and the action-effect principle (Prinz, 1997). The common-coding principle is that late perceptual products (event code) and early action precursors (action code) share a common representational domain. The action-effect principle is that actions are planned and controlled in terms of their effects—the events they cause to occur in the environment. The representation of the effect of an action is key to the successful unfolding of the action. The right panel of Figure 3 shows the two principles at work in performing a complicated coordination. A relative novice can produce a polyrhythm, for example, 3:4, under instructions to move his or her occluded hands so as to synchronize two visible motions. The design of the apparatus is such that the 1:1 elementary visible rhythm is achieved only when the hand motions are frequency locked at 3:4 (Mechsner et al., 2001). This research has a simple moral: to the extent that an intended effect corresponds to a simple detected event, the realization of the intent will be simplified.

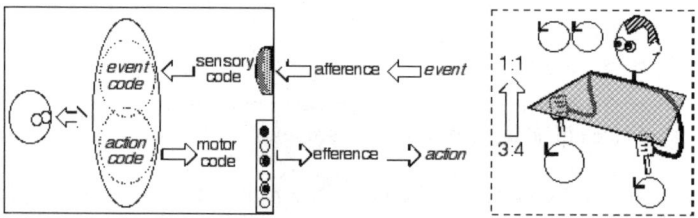

Figure 3. A further perspective on how the perception-action divide is crossed. Here, perception and action are taken to be partially commensurable (see text)

Theories directed at crossing the divide assume, of course, the validity of the reasons for the divide depicted in Figure 1. Many who study perception and action would probably agree that the reasons ought not to be regarded as ultimate truths.

Nonetheless, it is certainly the case that over the centuries scientific inquiry into perception and action has come to regard them, and their implications, as sacrosanct.

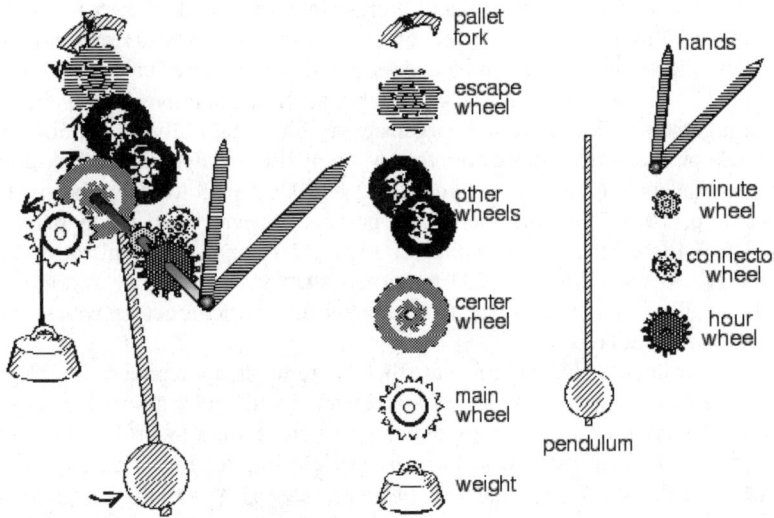

Figure 4. A pendulum clock provides a useful image of the mechanical-like anatomical units suggested by the Cartesian machine metaphor (see text). (Adapted from Gibbons G (1979) *Clocks and how they go*. Harper Collins, New York)

Let us focus on the alternative strategy: dissolving the divide. Minimally, two steps are required. The first step counters the conceptions portrayed in Figure 1. It is in the taking of this first step that the circle of ideas traced by Rosen in his paper of 1987 (and subsequently) must be entered. The second step is that of confronting the kind of theory of perception-action in need of development once the new conceptions are in place and the divide no longer exists. It is in the taking of the second step that the notions advanced in the 1987 paper by Schöner and Kelso must be brought to the fore and given careful examination.

3. Dissolving the perception-action divide: I. Units of function not units of anatomy

The divide, as Figure 1 highlights, is strongly motivated by anatomical distinctions situated within explanation by linear causal chains. The anatomical parts subserving the train of events from world to brain are not the same, and possibly categorically distinct from, those subserving the train of events from brain to world. To the degree that these event trains can be equated, respectively, with perception and action, as Figure 1 suggests they should, then the functions of perception and action can be ascribed to components of anatomy. An emphasis on anatomy in the explanation of the functional properties of perception and action is a natural emphasis given the Cartesian machine metaphor that has been the mainstay of theorizing on biological matters for more than 350 years. Descartes advocated that we explain all physical things (inanimate and animate) in the way machines or automata (such as the clock shown in Figure 4) are explained, through the properties of their independent parts. On animate functions Descartes commented: "I should like you to consider that these functions follow from the mere arrangement of the machine's organs every bit as naturally as the movements of a clock or the other automaton follow from the arrangement of its counter weights and wheels (*Treatise on man*, Section 202; see Cottingham et al., 1985, Volume I, p. 108)." By this metaphor, perception parts and action parts—the anatomical units of the middle panel of Figure 1—are mechanical units. In his 1987 lecture Rosen (1988a, p. 23) remarked sternly that the machine metaphor "has above all estranged biology from the world of theoretical physics and its quest for 'universal laws.'"

Following Rosen (1991), we can say that anatomical components, in common with mechanical components, are characterized by the direction of entailment: function F is entailed by its component C. In the natural world, entailment is causality; in the formal world it is strict implication (Rosen, 1988a). Diagrammatically, an entailment can be expressed as $X \rightarrow Y$. For an anatomical unit, $C \rightarrow F$. The primary property of an entailment in logic is that it propagates "truth" hereditarily—Y inherits the "truth" of X (Rosen, 1991, 2000). If X is assumed to be "true" (whatever "true" might be in a given context), then it must likewise be the case that Y is true. Thus, in the case of an anatomical component, F is inherited from C analogous to the functions of the mechanical parts depicted in Figure 4. In the case of a functional component, C is tied to F with F, in turn, dependent on the larger system of which C is a part. This closed loop of entailment can be diagrammed as $F \rightarrow C \rightarrow F$ and it can be read as saying that truth propagates hereditarily in both directions. As the nest building example in Figure 5 suggests, C is inherited from F and, in a sense, F is inherited from C.

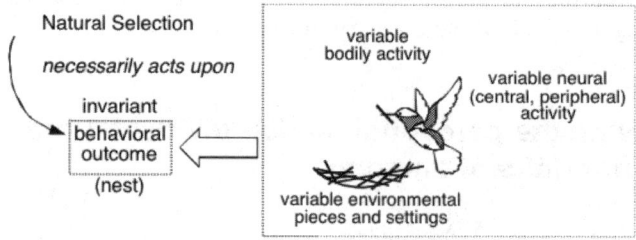

Figure 5. Dissolving the perception-action divide. Ia. Functional units rather than anatomical units (see text)

3.1. Functional entailment and impredicative definition

Entailments that are closed loop are *impredicativities* (a term coined by Poincaré; see Kline, 1980) and are counter to the *vicious circle principle* (Russell, 1903). An impredicative definition of a concept (or entity) is in terms of the totality to which it belongs as portrayed in Figure 6. The following definition of the football star Ronaldo is an example: "Ronaldo is the best striker on the Brazilian team." In this example the impredicative loop is that one must know the larger system (the totality of Ronaldo plus team mates) to characterize the smaller system (Ronaldo) but one cannot know the larger system in the absence of a characterization of the smaller system. The vicious circle principle is intended to filter out impredicativities. It disallows any whole that may contain parts that are definable only in terms of that whole. In system terms, it disallows consulting a larger system (a context) when attempting to understand a given system; only simpler subsystems can be invoked, in particular those whose defining properties and behaviors are context-free. The assumption motivating the principle is that any definition or description of a thing in a collection is suspect—not "objective"—if it relies on the use of the collection itself.

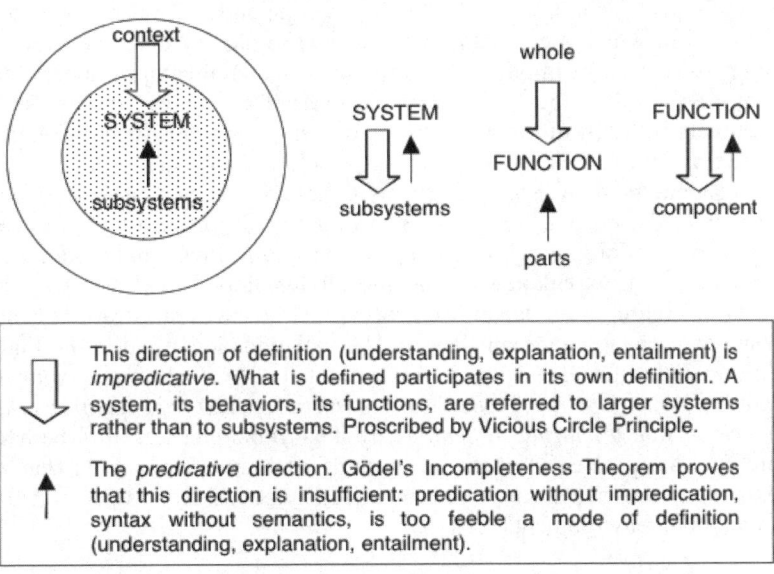

Figure 6. Dissolving the perception-action divide. Ib. Impredicative definitions beyond predicative definitions (see text)

The immediate goal of the vicious circle principle was to proscribe self-reference in the descriptions and definitions of systems and hence (given that referencing self is an instance of referencing in general) to proscribe semantics. In so doing, argued Russell (1903), all the paradoxes[1] that plagued logic and mathematics could be kept at bay. Consistent (paradox free) formalisms, could be assured, Russell suggested, by replacing things with impredicative properties by things with predicative properties (those true of individual things only). To confront the vicious circle principle and its far-reaching consequences fully we would need to review the conceptual developments from Russell to Turing and the contemporary interpretations of perception and action as computational processes. For our present purposes it suffices for us to defer to Rosen (1991, 2000). The negative lesson of Gödel's Incompleteness Theorem is that any attempt to formalize Number Theory, to replace its semantics by syntax in the manner suggested by Russell, Poincaré and Hilbert, will inevitably lose most of the truths of Number Theory. The positive lesson of Gödel's theorem is that impredicativities are generic—most systems (such as perception-action systems) are complex, possessing inherent impredicative loops. In his 1987 lecture Rosen was blunt in his summary of these matters (Rosen, 1988a, p. 27): "Complexity in our sense means that there is an essential semantic component to material reality that cannot be ignored."

We can look at the systems depicted in Figures 4 and 5 in terms of predicative and impredicative definition. The assumption from the Cartesian machine metaphor that a biological function F is entailed by its component C means minimally two things, both of which are suitably depicted by the old-fashioned pendulum clock and its parts. First, F is implicated strictly by the material composition and form of C. Any part of the clock in Figure 4 possesses a function F that is defined predicatively in terms of the part's properties, not impredicatively in reference to the larger system, the clock. Second, C is absolute. Its existence does not depend on the machine being operative and it continues to exist as a circumscribed entity when extracted from the machine, as the rightward portion of Figure 4 clearly shows. The situation is very different in the case of a functional system in which any component C is entailed by the overall function F. The F 'nest-building' depicted in Figure 5 is a dynamical process that achieves an invariant behavioral outcome over changing circumstances. The variations identified in the Figure are specific to F and generated by Cs that are entailed by F. *In sharp contrast with mechanical units, functional units are contingent (context dependent) and have no existence outside the larger system.* There is an important lesson to be had from Figure 5: Nature selects functional systems with functional units that do not necessarily abide scientifically convenient demarcations such as perception/action and brain/body/environment.

---

[1] The most well known non-mathematical paradox is the liar paradox. Consider statements such as "This sentence is false" or "I am lying." Denote the former statement as S. If S is true, then what it asserts is true, and so S is false. If S is false, then this is what it asserts, and so S is true. A famous mathematical paradox is Russell's paradox of classes. Consider N, the class of all classes that do not belong to themselves. (A class of ideas is an idea but a class of books is not a book.) Where does N belong? If N belongs to N, it should not by the definition of N. If N does not belong to N, it should by the definition of N (Kline, 1980).

4. Dissolving the perception-action divide: II. Causal organization rather than causal lineage

The upper panel of Figure 7 might be labeled "The fiction of the linear causal chain." Isolating a single line of production (solid arrows) in what is, factually, a dense network of causal entailments (all arrows) is an unavoidable scientific strategy. It means, however, that the chain is questionable as an ontological entity. At best the linear causal chain is a useful fiction. It is a fiction, however, that is more befitting simple systems that lack impredicative loops than complex systems that abound in impredicative loops. Rosen (2000) argued that science has mistakenly equated the notion of 'objectivity' with *mechanism* (that is, predicative, syntactic, chain-metaphor explanation) when, in fact, objectivity is to be equated with *complexity* (impredicative, semantic, support-metaphor explanation). Mechanism is simply a limiting case of complexity.

Hanson's (1969) deliberations on linear causal chains are most insightful. For him (p. 52), causes are "less like the links of a chain and more like the legs of a table…." The linear causal chain gives the false impression that the explanatory

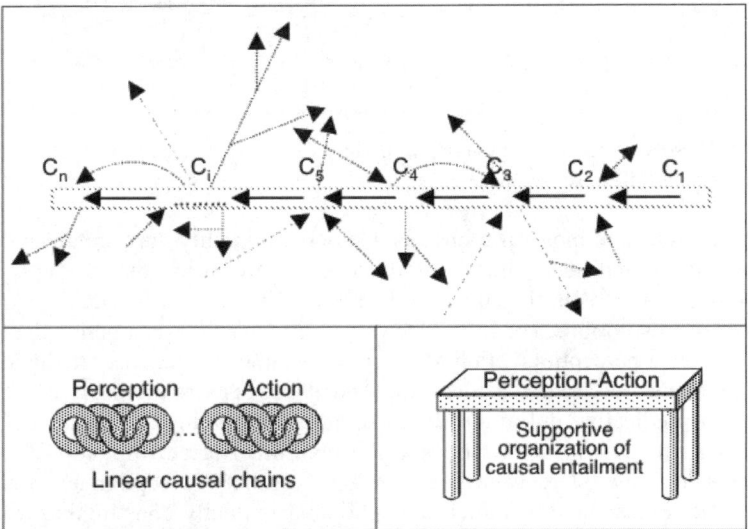

Figure 7. Dissolving the perception-action divide. II. Causal organization rather than causal lineage (see text). (Upper Figure is adapted from Figure 14 in Bunge M (1979) *Causality and modern science*. Dover, New York)

terms, the *causes*, and the things to be explained, the *effects*, are at the same logical level. To the contrary, the explanatory language is multi-leveled (Hanson, 1969). The 'cause x' and the 'effect y' do not name links in a chain of events but point to interlocked conceptual structures that place guarantees on the inference from cause to effect. A *support* metaphor is preferable rather than a *chain* metaphor. In Rosen's (1991, 2000) terms, *support* involves multiple modes of entailment, whereas *chain* is strictly and solely the entailment of subsequent states by present states. The two metaphors are picturesquely depicted in the lower panels of Figure 7.

5. Dissolving the perception-action divide. III. The perception-action cycle

Figure 8 identifies the Cartesian reflex arc: a linear causal linkage between sensory and motor processes and, on elaboration, between perception and action. Figure 8 also identifies the implications of Dewey's (1896) famous, incisive, yet historically ignored, critique. "The reflex arc is defective, in that it assumes

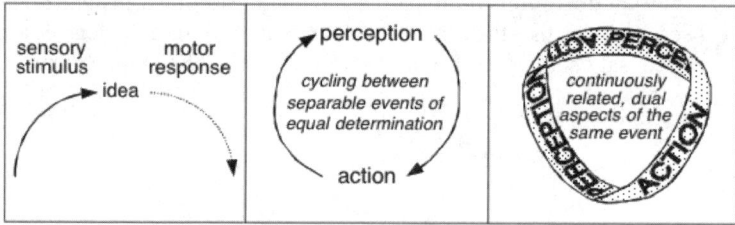

Figure 8. Dissolving the perception-action divide. IIIa. Perception-action cycle or Möbius band rather than reflex arc (see text)

sensory stimulus and motor response as distinct... in reality they are always inside a coordination and have their significance purely from the part played in maintaining or reconstituting the coordination...The arc...is virtually a circuit, a continual reconstitution...(p. 139)." Dewey's critique suggests a perception-action cycle, or even a perception-action Möbius band, rather than an arc. It suggests the metaphor of support rather than chain. And it suggests impredicative definition rather than predicative definition. In close agreement with my intellectual great-grandfather (my advisor's advisor received his Ph.D degree from John Dewey) I joined Robert Shaw two decades ago in suggesting that "Perception and action are of the same logical kind, symmetric, cyclic, and mutually constraining (Shaw & Turvey, 1980, p. 95)."

Figure 9 shows three related perspectives on the cycle formulation. The top panel expresses the perception-action cycle as a continual relation between perception degrees of constraint (*dc*) and action degrees of freedom (*df*), following ideas of Fowler and Turvey (1978), Fitch and Turvey (1978) and Turvey et al. (1978).

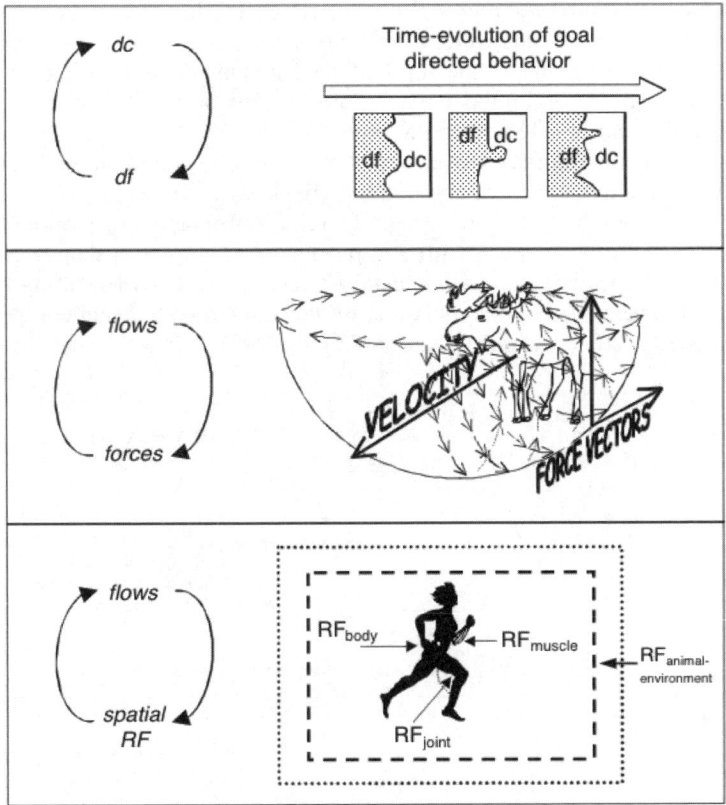

Figure 9. Dissolving the perception-action divide. IIIb. Three related perspectives on the perception-action cycle (see text)

The cycle is closed in terms of complementarity of *dc* and *df* and open in terms of kinds of *dc* and *df*. The middle panel shows the perception-action cycle as a continual relation between *flows* and *forces*, following ideas of Kugler and Turvey (1987). The forces produced by an animal move the animal and engender transformations of the optic array at the animal's point of observation. The transformations, often modeled as a velocity vector field, are referred to as optic flow (Gibson, 1966, 1979). The cycle is closed in terms of complementary forces and flows and is open in terms of the time-evolution of flow variables qua information. The bottom panel depicts how the perception-action cycle of Kugler and Turvey can be reconceived as a continual relation between *flows* and *spatial reference frames*. The idea that forces arise from intentional and perceptual adjustments in the origins of reference frames (RFs) nested at several length and time scales is Feldman's (1996; see present volume).

Figure 10 portrays two hypotheses about the perception-action cycle framed in terms of non-linear dynamics. One is the hypothesis of *complementary information and non-autonomous dynamics* (Beek et al., 1992). The other is the hypothesis of *emergent attractors in perception-action cycles* (Warren, 1998). The upper panel suggests that at the level of the movements composing an activity (such as the cascade juggle) the dynamics are non-autonomous. Systematic forcing contributions occur at select temporal points. At the level of the perception-action cycle, however, forcing may prove to be a function of information, not clock time. At this level, the dynamics are autonomous (Beek et al., 1992). In the lower panel the agent and environment are circularly related through information (e.g., as contained in the optical flow field) and mechanics (forces that change either the agent's relative location or environmental objects or both). The resulting behavior is emergent, determined by the dynamics of the coupled system and corresponding to the coupled system's attractor states (Warren, 1998).

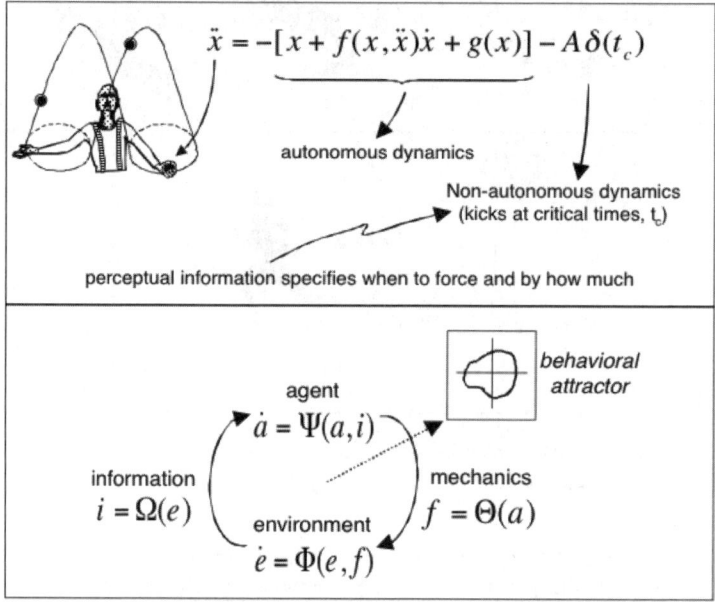

Figure 10. Dissolving the perception-action divide. IIIc. Non-linear dynamical perspectives on the perception-action cycle (see text)

6. Perception-action without the divide: A thoroughgoing coordination dynamics

Continual seems to be the appropriate adjective for the notion of a perception-action cycle. A *continual* process is one that goes on repeatedly. Applied to perception and action, continual means that perception is intermittent and action is intermittent with moments of non-perception and non-action. In contrast, *continuous* means unending and without interruption (Wilson, 1993). It seems more appropriate for the perception-action Möbius band. Continuous perception and continuous action would not cease for a second as the Möbius band of Figure 8 suggests. Whereas *continual* still echoes the perception-action divide (perception-action-perception-action…and so on), *continuous* does not. In brief, to dissolve the divide in a truly thoroughgoing manner, we need to make a theoretical advance beyond descriptions in terms of continual states. We need to find a way to express perception and action as concurrently continuous. Stated more pragmatically, we need to comprehend how variables that convenience and convention assign to the perception side of the divide, and variables that convenience and convention assign to the action side of the divide, can be absorbed by one or a few variables that have no side. Roughly speaking, it was this latter comprehension, I think, that Schöner and Kelso were seeking in their 1987 lecture.

The key idea is order parameter or collective variable. This is a measurable quantity that expresses a coherent relation among parts and processes. It is, of course, a central concept in Haken's (1977, 1983) synergetics. It is also a concept that is defined impredicatively: collective variable → parts → collective variable. In impredicative terms, Kelso (1995, p. 145) refers to order parameters as " *semantic*, relational quantities that are intrinsically meaningful to system functioning" and their dynamics as "context dependent" (and at odds, therefore, with "ordinary physics").

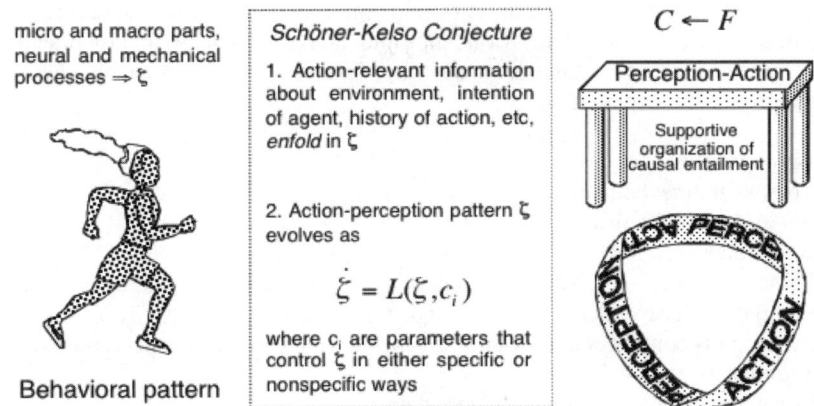

Figure 11. Perception-action without the divide realized by a thoroughgoing coordination dynamics

In the left panel of Figure 11 I have supposed, for purposes of illustration, that there is a collective variable for the behavioral pattern of running. Let this variable be designated ζ. Then, the *Schöner-Kelso Conjecture* (as I like to refer to their theoretical proposal of 1987) is that each of the various influences on the behavioral pattern can be interpreted (modeled) in terms of the very same quantity that expresses the behavioral pattern. That is, all the various influences *enfold* in ζ. The time-evolution of the behavioral pattern subject to multiple influences can then be captured, in principle, by a single first-order equation in ζ —as noted in the middle panel of Figure 11. In this formulation, perception and action entwine as a Möbius band, the support metaphor rather than chain metaphor of causal entailment holds sway, the units are functional and context-dependent rather than anatomical and context-independent,[2] and impredicative definitions are in ascendency—as intimated in the right panel of Figure 11.

The Schöner-Kelso Conjecture is a natural extension of the basic explanatory framework that is dynamical systems theory. In this framework, a system's behavior is explained by identifying one or a few collective variables that capture the qualitatively distinct patterns exhibited by the system as it unfolds over time and by describing these patterns (their emergence and dissolution) in the precise terms of attractors, bifurcations multistability, and related notions. The strategy of folding together all aspects within the dynamics of the collective variable is an effort to embrace the full complexity of biological perception-action without a proliferation of arbitrary divisions.

7. Perception-action without the divide: Conceptual barriers

Reluctance to pursue an explanation of perception-action without the divide and in strictly dynamical terms stems from multiple sources. I list the major ones here. They are not fully independent but the nuances that distinguish them are illuminating. The explanation:

1. is not visualizable.
2. provides no mechanism.
3. ignores biomechanics.
4. cannot be simulated.
5. glosses over obvious natural boundaries (perception/action, brain/body/environment).
6. provides no context-independent parts with which to build a perception-action system.
7. does not presume a fixed set of state variables.
8. relegates explicit material processes to abstractions (of enfolded, collective-variable dynamics).

---

[2] A source of nice examples is Schöner et al's (1998) investigation of postural control. "physiological or biomechanical subsystems are not in themselves separable units, even though they are often analyzed in this manner. Multiple subsystems may be recruited to interact as single units in terms of the elementary behaviors (p. 327)." "The control system for posture cannot be separated into sensory and motor components (p. 333)."

In the history of science a willingness to renounce explanations in literal and sensuous[3] terms has developed slowly and with much opposition (Cassirer, 1950). It began and reached fruition first in theoretical physics. The challenges of quantum theory provided the major impetus. For an early renunciation we may consider Heisenberg on the atom (quoted by Cassirer, 1950, p. 117): "*All* its qualities are inferential: no material properties can be directly attributed to it. That is to say, any picture of the atom that our imagination is able to invent is for that very reason defective. An understanding of the atomic world in that primary sensuous fashion...is impossible."

Not surprisingly, a reluctance to abandon visualization is conjoined with a reluctance to abandon mechanism. Historically, to explain a phenomenon within the Cartesian machine metaphor is to give an account that is machine-like. This is what it means to provide a *mechanism* for the phenomenon. A mature mechanistic account would, of course, be expressed mathematically. In such an account, however, the traditional role of the mathematical symbols has been to represent the components and events in the particular machine that is proposed as the phenomenon's simulacrum. In a word, the symbols must function as *replicas* of the details and properties of the simple machines whose summed effects constitute the phenomenon. The prominent Victorian physicist William Thompson (Lord Kelvin) wrote (Cassirer, 1950, p. 115): "I am never content until I have constructed a mechanical model of the object that I am studying. If I succeed in making one, I understand; otherwise, I do not." Thompson's words expressed the desire, common to the physicists of his time, to make things plain, to express them in a sensible form, a form that is easily visualized. To this end, the symbols in a mechanism's mathematical formalism must connect to the hypothesized material objects composing the mechanism. Eddington (1928/1958) summarized the likes of Thompson in these terms: "It was the boast of the Victorian physicist that he would not claim to understand a thing until he could make a model of it; and by a model he meant something constructed of levers, gears, wheels, squirts, or other appliances familiar to an engineer. Nature in building the universe was supposed to be dependent on just the same kind of resources as any human mechanic; and when the physicist sought an explanation of phenomena his ear was straining to catch the hum of machinery. The man who could make gravitation out of cogwheels would have been a hero in the Victorian age."

What Thompson desired for an explanation (and what Eddington derided) was a *simulator*. Thompson was convinced that a mechanical simulation led to an inherent understanding of any given physical phenomenon. Eddington on the other hand saw mechanical simulations as possibly aiding visualization of a phenomenon but offering no guarantees of improved understanding. For Eddington, enhanced understanding came at the price of picture-ability. And it came at the price of concreteness. The symbols of a mathematical model for

---

[3] The usage of *sensuous* here is its first meaning: of or relating to the senses or sensible objects; suggesting pictures or images of sense. It is likewise the case for the usage of *sensible*: capable of being perceived by the senses; apprehensible through the sense organs.

Eddington referred to abstractions rather than to hopefully visualizable machine parts. The lesson of quantum physics shaping Eddington's perspective, a lesson that is repeated by dynamical systems theory, is that all that can be legitimately expected of a (lawful) explanation is that it provide correct predictions. Whether or not the process of setting up such predictions yields a pictorial representation of the situation is of no consequence (March & Freeman, 1963).

The contrast between Thompson and Eddington will not be unfamiliar to the modern observer of the sciences of perception and action. Indeed, to such an observer it should be fairly apparent that scientists with Thompson's sentiments compose the majority with Eddington-like scientists few and far between. The prevailing attitude is that a sensory or motor phenomenon can be truly understood only if it can be pictured, that is, only if it can be ascribed a mechanism. Such is the evolutionary stage of inquiry into perception and action.

In part, I chose the quotations of Thompson and Eddington because they lead us without complication to a fundamental theme in Rosen's 1987 lecture: the identity between the concepts of simulation and mechanism. This identity is the convergence of two 17th century ideas—Hobbes's (1651/1968) idea that thinking consists of symbolic operations on special brain tokens and Newton's idea of a partition between states (that belong to a system) and dynamical laws (that inhere in the system's environment). Explicit recognition of Newton's partition and its implications has been one of Rosen's major contributions.

The contemporary manifestation of Hobbes's idea is the Turing machine, a mathematical machine where the parts are symbols that can come in different sizes (strings of symbols) and shapes (syntactic roles) and exhibit different motions (transformations by rule). In a Turing machine, meaningless symbols are analogous to particles in Newton's framework and symbol strings are analogous to configurations of such particles. Further, in a Turing machine, the manipulation of symbols by fixed rules (of arithmetic) that are external to the symbols is analogous to the changes in the states of particles by laws (of forces) that reside in the environment of the particles. At root, the mathematical machine and the Newtonian formalism for material systems are identical (Rosen, 1988a, 1998b, 1991, 2000). Recursiveness is the core of both that is, for both, conversion of the present to the immediate future *exhausts* their entailment structure.

For present purposes, the significance of the mathematical machine is that it is a way of thinking about the syntactic aspect of mathematics separately from its semantic aspect. The syntactic aspect is the rules that compose the mathematics' internal grammar. The semantic aspect is the external matters that the mathematics refers to, that is, what it is about or what it means. We recall the impredicativities or self-referring definitions and Russell's and Poincaré's desire to prohibit their occurrences in logical and mathematical systems. Those who held similar views, especially the mathematical polyglot Hilbert, believed that the desired prohibition (the vicious circle principle) could be implemented by restricting such systems to their syntactical aspect and adding further syntactic rules as needed to replace the semantic features. Any mathematical system (e.g., Euclidean geometry) stripped of its external referents and encoded strictly by syntactic rules is designated "formalizable." A mathematical system that can be formalized is equivalent, therefore, to the strictly syntactical (by definition) mathematical machine. The primary consequence of this equivalence is that a machine can "do" the mathematical formalism. On extension, this nexus of ideas leads to the following conclusion (Rosen, 1991, p. 185): "Simulation is what machines do, and a system (formal or material) *is* a machine if it simulates or can simulate, something else."

The equations of 'machine' with 'simulator', 'mechanism' with 'simulation', and 'simulation' with 'strictly syntactical' imply that a material system whose phenomena can not be simulated is a material system that is not a machine. Eddington's chiding of Thompson and his fellow Victorian scientists may be interpreted as a reaction against the belief that the natural world can be rendered fully in a strictly syntactic form. In his 1987 lecture Rosen (1988, p. 27) stated this possible interpretation more firmly: "Purely syntactical models of material nature are too impoverished in entailment to mirror the causal sequences which can occur in nature...."

Despite appearances, the foregoing paragraphs are not meant to be pedantic. The historical references made to formalism and the centrality of recursion in Newton's mechanics and Turing's machines are necessary to bring out a key reason for reluctance in regard to dissolving the perception-action divide in the manner suggested by Figure 11. The kind of explanation that Schöner and Kelso championed in 1987 is not translatable into a picture-able machine. It is not simulable in Thompson's terms. But it is not simulable in a deeper sense. It is not simulable in Rosen's terms, those he sketched in his lecture of 1987. Systems with inherent and irremovable impredicativities can not, by definition, be simulated. Such systems, as observed above, are complex. The systems that can be simulated are simple.

Figure 12 is an example of the *modeling relation*. Generically, the relation comprises two paths: (1) and (2) + (3) + (4). Path (1) is causal entailment in the natural system. In the second path, (2) encodes phenomena in the natural system to propositions in the formal system, (3) generates theorems entailed strictly by the propositions qua hypotheses, (4) decodes theorems as predictions about the natural system. The formal system is a model of the natural system if the same consequences (answers) are reachable by either path. In Figure 12 the formal system is Newton's recursion rule (as shown) applied to the system's phase at t_0. The modeling relation may be viewed as an embodiment of the notion of *natural law*. Although Newton's model of natural law expressed in Figure 12 has had great success in dealing with material phenomena, it is inadequate, by the arguments developed above, for complex systems.

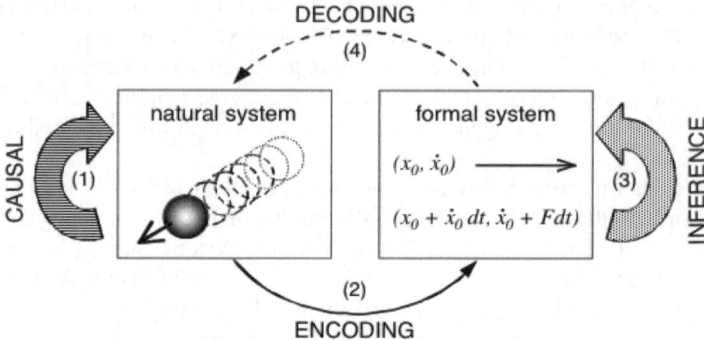

Figure 12. The poverty of entailment. This modeling relation limited to syntactic, Newtonian entailment is insufficient for division-less perception-action interpreted as a thoroughgoing coordination dynamics. [Adapted from Figure 1 in Rosen (1988a)]

In sum, in their lectures of 1987, Schöner and Kelso showed us a novel, dynamics framework within which the perception-action divide dissolves and Rosen showed us the nature of the primary conceptual barrier that must be overcome for such a framework to mature. We should call this barrier *the problem of impoverished entailment*.

Acknowledgments

Preparation of this manuscript was supported by grant SBR 00-04097 from the National Science Foundation. The contributions of Claudia Carello are too numerous to mention.

References

Beek P, Turvey MT, Schmidt RC (1992) Autonomous and nonautonomous dynamics of coordinated rhythmic movements. Ecol Psychol 4, 65-96

Cassirer E (1950) The problem of knowledge. Yale University Press, New Haven, CT

Cottingham J, Stoothoff R, Murdoch D (1985) The philosophical writings of Descartes. Cambridge University Press, Cambridge

Dewey J (1896) The reflex arc concept in psychology. Psychol Rev 3, 357-370

Eddington A (1928/1958) The nature of the physical world. University of Michigan Press, Ann Arbor, MI

Feldman A (1996) Spatial frames of reference for motor control. In: Latash M (ed.) Progress in motor control, Vol I, Bernstein's traditions in movement studies. Human Kinetics, Champaign, IL, 289-314

Fitch H, Turvey MT (1978) On the control of activity: Some remarks from an ecological point of view. In: Landers D, Christina RW (eds.) Psychology of motor behavior and sport. Human Kinetics, Champaign, IL, 3-35

Fowler CA, Turvey MT (1978) Skill acquisition: An event approach with special reference to searching for the optimum of a function of several variables. In: Stelmach G (ed.) Information processing in motor learning and control. Academic Press, New York, 2-40

Gibson JJ (1966) The senses considered as perceptual systems. Houghton-Mifflin, Boston

Gibson JJ (1979) The ecological approach to visual perception. Houghton-Mifflin, Boston

Haken H (1977) Synergetics. Berlin, Springer Verlag

Haken H (1983) Advanced synergetics. Berlin, Springer Verlag

Hanson NR (1969) Patterns of discovery. Cambridge University Press, Cambridge

Hobbes T (1651/1968) Leviathan. Penguin Books, London

Kelso JAS (1995) Dynamic patterns. MIT Press, Cambridge MA

Kline M (1980) Mathematics: The loss of certainty. Oxford University Press, New York

Kugler PN, Turvey MT (1987) Information, natural law and the self-assembly of rhythmic movement. Erlbaum, Hillsdale NJ

March A, Freeman IM (1963) The new world of physics. Vintage Books, New York

Mechsner F, Kerzel D, Knoblich G, Prinz W (2001) Perceptual basis of bimanual coordination. Nature 414, 69-73

Pellionisz A (1985) Tensor network theory of the central nervous system and sensorimotor modeling. In: Palm G, Aertson A (eds.) Brain theory. Springer Verlag, Berlin

Prinz W (1997) Perception and action planning. Eur J Cog Psychol 9, 129-154

Rosen R (1988a) The epistemology of complexity. In: Kelso JAS, Mandell A, Shlesinger MF (eds.) Dynamic patterns in complex systems. World Scientific, Singapore, 7-29

Rosen R (1988b) Similarity and dissimilarity: A partial overview. Hum Movement Sci 7, 131-154

Rosen R (1991) Life itself. Columbia University Press, New York

Rosen R (2000) Essays on life itself. Columbia University Press, New York

Russell B (1903) Principles of mathematics. Cambridge University Press, Cambridge

Schöner G, Kelso JAS (1988) Dynamical patterns of biological coordination: Theoretical strategy and new results. In: Kelso JAS, Mandell A, Shlesinger MF (eds.) Dynamic patterns in complex systems. World Scientific, Singapore, 77-102

Schöner G, Dijkstra TMH, Jeka JJ (1998) Action-perception patterns emerge from coupling and adaptation, Ecol Psychol 10, 323-346

Shaw RE, Turvey MT (1980) Methodological realism. Behav Brain Sci 3, 94-96

Turvey MT, Shaw RE, Mace W (1978) Issues in the theory of action: Degrees of freedom, coordinative structures and coalitions. In: Requin J (ed.) Attention and performance VII. Erlbaum, Hillsdale, NJ, 557-595

Warren WH (1998) Visually controlled locomotion: 40 years later. Ecol Psychol 10, 177-220

Wilson K (1993) The Columbia guide to standard American English. Columbia University Press, New York

PART II: COGNITIVE CONTRIBUTIONS TO COORDINATION DYNAMICS: ATTENTION, INTENTION AND LEARNING

A Dynamical Approach to the Interplay of Attention and Bimanual Coordination

Jean-Jacques Temprado

UMR 6152 "Mouvement et Perception" Université de la Méditerranée. Faculté des Sciences du Sport, Marseille – France

Despite their common origin, studies on motor coordination and on attentional load have developed into separate fields of investigation, bringing out findings, methods, and theories which are diverse if not mutually exclusive. Sitting at the intersection of these two fields, this article addresses the issue of behavioral flexibility by investigating how intention modifies the stability of existing patterns of coordination between moving limbs. It addresses the issue, largely ignored until now, of the attentional cost incurred by the Central Nervous System (CNS) in maintaining a coordination pattern at a given level of stability, in particular under different attentional priority requirements. The experimental paradigm adopted in these studies provides an original mix of a classical measure of attentional load, namely, Reaction Time (RT), and of a dynamic approach to coordination, most suitable for characterizing the dynamic properties of coordinated behavior and behavioral change. Findings showed that central cost and pattern stability covary, suggesting that bimanual coordination and the attentional activity of the CNS involved in maintain such a coordination bear on the same underlying dynamics. Such a conclusion provides a strong support to a unified approach to coordination encompassing a conceptualization in terms of information processing and another, more recent framework rooted in self-organization theories and dynamical systems models

One of the most amazing capabilities of biological systems is their ability to alter behavior on several time scales in response to influences of various origins (environmental, volitional, attentional, etc...). A main challenge for movement scientists is to identify the mechanisms and principles through which behavioral flexibility arises from the interplay between constraints inherent to the behaving system and extraneous influences. Recently, much attention has been drawn on rhythmic interlimb coordination, particularly from a dynamical systems approach (see Kelso, 1995, for a review). At the core of this approach is the recognition that coordination states -- which can be captured by informational quantities such as the phase relation between components -- result from a self-orgnazied process and "emerge" from the mutual influences between interacting components (e.g., neural, muscular, mechanical, energetic). Numerous experiments on different kinds of motor coordination have shown that the salient features of the coordination dynamics can often be elucidated by driving the system toward a point of instability in which a qualitative change or phase transition occurs. Such a *phase transition* is characterized by an abrupt switch of the relative phase value and it is preceded by an increase of relative phase variability of previously stable patterns. Thus, the concepts of stability and flexibility are central to consideration of coordination dynamics. A significant step in efforts to understand stability and flexibility of coordination dynamics was to open the dynamic approach to coordination to cognition, that is to intentional, attentional, perceptual and memory influences (e.g., Scholz and Kelso, 1990; Wuyts et al., 1996; Zanone and Kelso, 1992, 1997). These specific influences may be conceptualized as additional forces acting on the collective variable dynamics, perturbing the system toward a new required pattern or altering the stability of existing patterns (Schöner and Kelso, 1988a,b).

This paper addresses the issue of behavioral flexibility by investigating how intention and attention alter the stability of spontaneous coordination dynamics between oscillating limbs. Moreover, it tackles the issue of the attentional cost incurred by the nervous system (CNS) in maintaining a coordination pattern at a given level of stability. We go on to provide empirical support for the prediction that attention resides within the self-organizing dynamics of behavior, forming an indissociable part of such dynamics. As a result, we show that attention and coordination dynamics mutually interact in a circular causality: Attention constrains coordination dynamics but coordination dynamics also constrains attention.

1. The HKB dynamic model of bimanual coordination

Bimanual coordination is characterized by two preferred patterns of coordination -- in-phase and anti-phase (Kelso, 1984). In the case of motion in the horizontal plane, the in-phase pattern involves symmetric motion of the hands in opposite directions, due to the simultaneous activation of homologous muscles, whereas the anti-phase pattern involves motion in the same direction, with simultaneous activity of the antagonist muscles. The in-phase pattern proves to be more stable than the anti-phase pattern, and an unavoidable switch from the latter to the former occurs when oscillation frequency increases beyond a given critical value. These phenomena are exhibited in a variety of tasks and situations, suggesting that the same coordination principles are harnessed by very different ensembles of components across multiple levels of description (see Kelso, 1995).

The behavior exhibited spontaneously by bimanual coordination was formalized by Haken, Kelso and Bunz (1985) (henceforth, HKB model) through the dynamics of the relative phase between the limbs, seen as nonlinear oscillators linked through a low-energy (nonlinear) coupling function (Schöner et al., 1986). Intuitively, such dynamics may be rendered as a potential landscape composed of two valleys of different depths, in which an overdamped particle, representing the current coordination state of the system, would move freely following a gravitational force (see Figure 1).

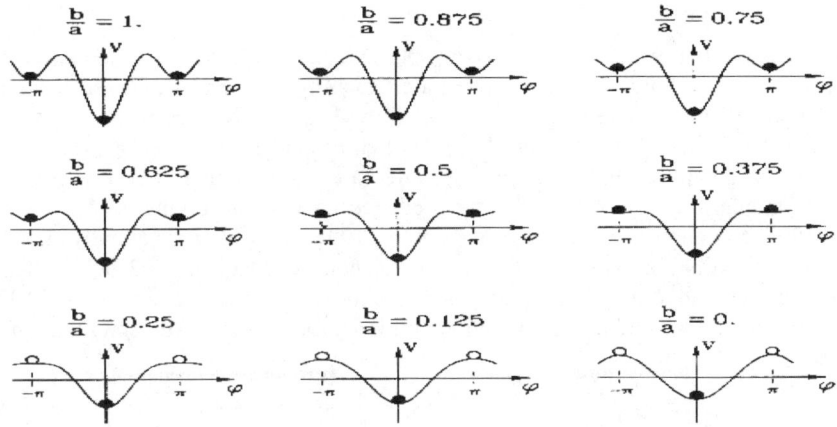

Figure 1. The potential function equation developed by Haken, Kelso and Bunz (1985) - $V(\phi) = -a \cos \phi - b \cos 2\phi$ - on the interval [-180°, 180°] for different values of b/a. In its original form the HKB equation lacked a stochastic force (added by Schöner et al., 1986) and a symmetry breaking term (added by Kelso et al., 1990). The ratio b/a represents the oscillation frequency effect on the potential landscape. As the ratio b/a is decreased (which corresponds to an increase in frequency), the anti-phase pattern becomes less stable. At a specific value of b/a, the local minimum at 180° disappeared entirely, and any small fluctuation kicks the system into the remaining minimum at 0°. Then, the potential landscape becomes monostable. This corresponds to the spontaneous switch from anti-phase to in phase observed experimentally (adapted from Kelso, 1995).

The curve presented in Figure 1 is characterized by a lower and a higher trough "located" at 0° (in-phase) and 180° (anti-phase), respectively. Thus, depending on its initial state, the system is ineluctably attracted to and eventually stabilizes into one of the two valleys, thereby defining two stable states or attractors of the coordination dynamics. The relative depths of the two valleys may vary as a function of a control parameter, which may represent in this particular instance oscillation frequency, that modifies the attractive strength of the stable states, that is, the "force" that they exert on relative phase toward in-phase or anti-phase. Fluctuations of relative phase thus quantify the stability of the various coordination patterns and determine whether and how their stability changes with scaling a control parameter. In particular, pattern switching is preceded by critical fluctuations, assessed as an increase in the standard deviation of the relative phase as oscillation frequency increases. These fluctuations subsequently decrease after the transition, once the in-phase pattern is adopted (Kelso et al., 1986). Assuming that the strength of the stochastic noise remains constant for all oscillation frequencies (Schöner et al., 1986), such an increase in relative phase fluctuation suggests a decrease in the coupling strength between the oscillating components. Such a loss in coupling strength is more salient for the intrinsically less stable anti-phase pattern than for the in-phase pattern.

2. Flexibility of spontaneous coordination dynamics

The constraints imposed by underlying spontaneous coordination dynamics on the actual behavior, however, do not preclude flexibility. Several studies have shown that subjects may intentionally: (a) switch between preferred coordination patterns (Carson et al., 1994; Carson et al., 1996; Scholz and Kelso, 1990); (b) delay or inhibit the spontaneous transition from one pattern to another (Lee et al., 1996); (c) momentarily stabilize an existing preferred coordination pattern (Amazeen et al., 1997; Swinnen et al., 1996; Wuyts et al., 1996); and (d) permanently stabilize a novel coordination pattern with learning (Zanone and Kelso, 1992, 1997). These flexible behaviors may be readily accounted for in the Haken et al. (1985) conceptualization. Intentional switching amounts to a selective (de)stabilization of

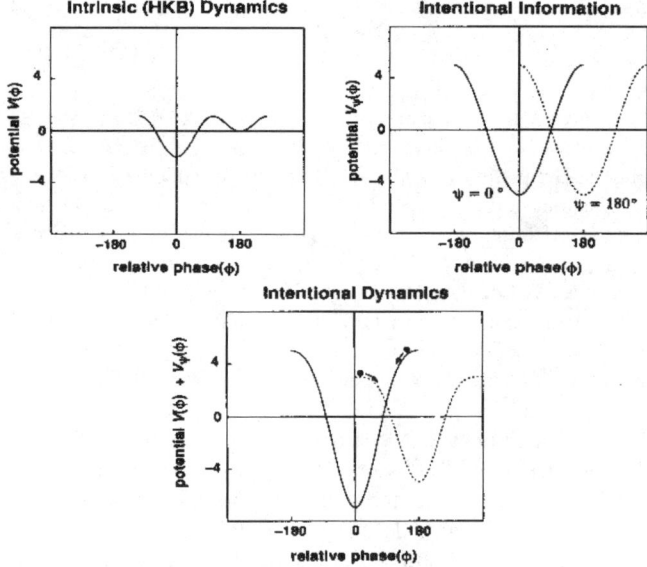

Figure 2. The potential function (equation) developed by Schöner & Kelso (1988a,b) - ($V(\Psi) = V(\phi) - c.\cos(\phi - \Psi)$). Behavioral dynamics result from the superposition of the spontaneous dynamics represented by $V(\phi)$ and the behavioral information represented by $V(\Psi) = -c.\cos(\phi - \Psi)$ (adapted from Kelso, 1995).

existing preferred coordinative states, whereas learning corresponds to the stabilization of a novel attractive state through practice. For both phenomena to be captured formally, an additional term is needed in the HKB model (Schöner and Kelso, 1988a, 1988b; henceforth, S and K model). Such an additional attractive force was coined as "behavioral information" (Schöner and Kelso, 1988a,b). Behavioral information may then compete or cooperate with the spontaneous coordination tendencies, depending on whether it coincides with the spontaneous stable states defined by the coordination dynamics, or not. Thus, the dynamic properties of the performed coordination pattern are determined by the location

and the strength of the additional term relative to those of the spontaneous coordination tendencies (Schöner and Kelso, 1988a,b; Schöner et al., 1992) (Figure 2.).
The present article expounds on the foregoing dynamical framework in further establishing the mapping between pattern stability and the CNS attentional activity through both behavioral and central cost measures.

3. Attentional influences on coordination dynamics

Attention has long been considered a prominent mediator, which allocates "energetic resources" (van der Molen, 1996; Sanders, 1998) or "effort" (Kahneman, 1973) to a given task. It is well known that allocation of resource is mediated by specific neural structures and networks. Indeed, attention involves a set of brain processes which interact mutually and with other brain processes in the performance of perceptual, cognitive and motor tasks (see Parasuraman, 1998, for an overview). Our assumption is that in the lack of a direct access to the activity of these attentional networks, one can assess indirectly the "effort" sustained by the central nervous sytem to perform a motor task using *attentional load* measure and a classical method in experimental psychology, that is the *dual-task paradigm*. In the dual task paradigm, subjects perform simultaneously a primary motor task and a secondary task (e.g., a RT task). The attentional priority given to each task is manipulated; they can be instructed either to share their attention between the two tasks or to direct their attention to one task or the other (e.g., Navon, 1990; Tsang et al., 1995; Tsang et al., 1996). When attention is focused on the primary task, the amount of attentional resources allocated to that task increases in order to maintain or increase performance. Such a supplemental allocation leads to a depletion in the resources devoted to the other, non-priority task. Then, performance in the primary task improves at a measurable cost for the non-priority task, manifesting a so-called "performance tradeoff ". The performance trade-off measures the difficulty for the CNS to perform the primary motor task, that is the amount of ressources allocated to this task (Navon, 1990).
In our research program we adopted a dual-task paradigm involving a rhythmic bimanual coordination task and a discrete Reaction Time (RT) task. From our dynamic perspective, the attentional cost measured by RT reflects the intentional force added to the spontaneous coordination dynamics -- as formalized by the Schöner and Kelso's model -- in order to maintain and/or to further stabilize a coordination pattern. Thus, RT and/or the difference in RT afford a valid index of the CNS attentional activity (the "effort") associated to maintaining bimanual preferred coordination patterns at a given level of stability. An originality of this procedure is that it establishes a correspondence between two conceptualizations of coordination, until now separated: The dynamical landscape, defined by the coordination dynamics and assessed through stability measures, and the information processing activity of the CNS, assessed through attentional load measures.

4. Stability of preferred patterns and central cost

In a first study (Temprado et al., 1999), we aimed to check whether maintaining preferred patterns of coordination (i.e., in-phase and anti-phase) occurs at an attentional cost, which may vary as a function of their proper stability. Other goals were (a) to test whether preferred bimanual coordination states might be stabilized intentionally through attentional focus directed to the coordination task and (b) to determine whether intentional stabilization of preferred bimanual patterns incurs a central cost. To verify these assumptions, we analyzed the performance tradeoff between RT and relative phase variability, as a function of several attentional priority conditions. We expected the attentional demands to be proportional to the initial stability of the coordination patterns and to the amount of stabilization actually effected.

Grabbing a joystick with each hand, participants were asked to perform periodic forearm pronation-supination movements at a freely chosen movement rate. They had to execute one of two coordination patterns between hand oscillations: (a) *in-phase* (0° of relative phase), in which homologous rotator muscles were recruited simultaneously, so that the joysticks would move in opposite directions; and (b) *anti-phase* (180° of relative phase) in which non-homologous muscles were activated, leading to joystick movements in the same direction. Participants performed both single task and dual-task conditions (for details, see Temprado *et al.*, 1999). In a dual-task condition, the coordination and the RT tasks were associated, with different attentional priority requirements: (a) a condition with attention shared between the bimanual task and the RT task; (b) a condition with priority given to the bimanual task; and (c) a condition with priority given to the RT task. Each condition was realized with both the in-phase and anti-phase patterns. To assess performance level in the bimanual task, we measured the produced relative phase and RT over each trial.

Comparison of relative phase and RT profiles affords a general picture of the results. For the three conditions, there was a significant difference in both relative phase and RT between anti-phase and in-phase patterns (Figure 3). The more stable pattern (in phase) was also the less costly to maintain, whatever the priority condition. The analysis concerned differential effects of the various attentional priorities in the dual-task conditions. Statistical analyses revealed that for the in-phase pattern, there was no evidence for any change as a function of conditions. In contrast, for the anti-phase pattern, focusing attention on RT led to an increase in relative phase fluctuations with respect to the condition in which attention was shared between the tasks. Giving a priority to bimanual coordination led to a decrease of such fluctuations with respect to the two other conditions. Then, a main finding is that stability of the coordination pattern is affected by manipulations of attentional priority, at least for the anti-phase pattern: Pattern fluctuations are minimized or increased when the attentional focus is set on the main or the additional task, respectively (Figure 3A). Considering RT (Figure 3B), irrespective of the bimanual pattern, focusing on the task improved performance, whereas focusing on an additional task decreased performance. Focus on the bimanual task led to the worst performance. Moreover, comparison of relative phase and RT profiles affords a general picture. Regardless of the pattern and the task, the shared-attention condition corresponded to an intermediate level of performance, as compared to the other two conditions, in which priority was given to one task or the other.

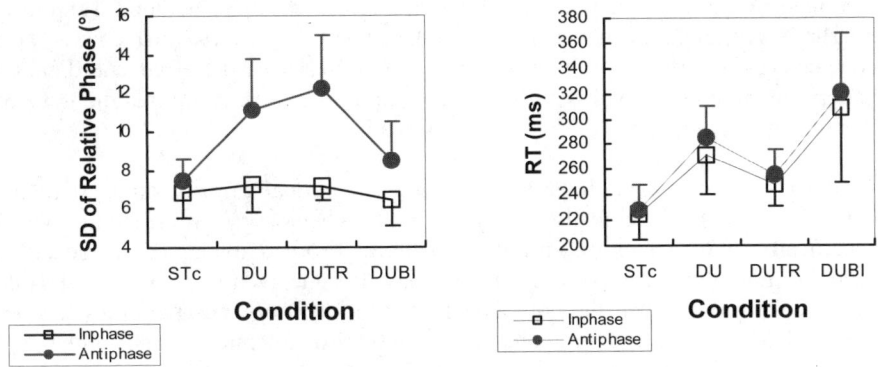

Figure 3. Performance in the bimanual task (Figure 3A). Figures 3A and B present the mean SD of performed relative phase and the mean RT, respectively, for the in- and anti-phase patterns (denoted by □ and, ● respectively), as a function of the experimental conditions. Data is averaged across trials and subjects. Mean RT as a function of the experimental conditions (Figure 3B). RT performance was faster in the situation where attention priority was set on the RT task than for the two other conditions.

These results suggest that, as expected: (1) maintaining preferred patterns of coordination (i.e., in-phase and anti-phase) occurs at an attentional cost, which vary as a function of their proper stability there; (2) there is a tradeoff between the bimanual and the RT tasks in the dual-task conditions, depending on the attentional focus. Note, however, that such a tradeoff does not affect the pattern stability of the in-phase pattern. Moreover, in all dual-task conditions, the stability that the in-phase pattern finally attained was still larger than that of the anti-phase pattern, attesting to the persisting influence of the differential stability characterizing the stable states of the coordination dynamics. Nonetheless, the decrease in variability for the in-phase pattern was smaller than for the anti-phase pattern when attention priority was given to the bimanual task, whereas the change in RT was equivalent for both patterns. This finding indicates that RT does not match precisely the stabilization actually realized. Although the most stable pattern (in-phase) was basically less costly in attentional demands when attention was shared, it was more demanding to further stabilize, suggesting a differential "resistance" of the coordination patterns that correlates with their baseline stability. Thus, coordination dynamics act as a constraint that limits the subject's intention to sustain and extra-stabilize coordination patterns (Scholz and Kelso, 1990). Our results suggest that this constraint not only limits the stabilization that these preferred patterns might assume, but also that the associated attentional cost is affected by their inherent stability.

To sum up, the results of the study by Temprado et al. (1999) assert to the role of the CNS, which can be assessed through RT measure, in maintaining a given level of pattern variability, even in spontaneous situations in which the level of stability exhibited by the coordination patterns is not specified in any manner. Therefore, interlimb coordination never occurs independent of central influences. This conclusion strongly questions a hierarchic view that sees preferred coordination patterns as running in total isolation from central processes, and raises the issue of the informational nature of inter-limb coordination.

At the conceptual level, intention - *via* the attentional priority directed to the collective variable - constitutes, by definition, behavioral information, which specifically stabilizes the coordination pattern. From a dynamical perspective, such an improvement in pattern stability may be conceived of as a strengthening of the coupling between the components due to intention and attention (Schöner and Kelso, 1988a,b). Thus, one can distinguish intentional and attentional contributions. Performing coordination patterns requires an explicit intention to be performed, taking the form of a specification of the relative phase of the pattern. On the other hand, attentional contribution can be conceived of as the tuning of the coupling strength between the components needed to perform the intended (i.e., explicitly specified) pattern. In other words, both intention and attention are needed to entail an informational coupling between the oscillating components. These intentional and attentional contributions may be represented by different parameters in the equation of the potential function of the Schöner and Kelso's model: Ψ is the relative phase specified by behavioral (intentional) information and c represents the coupling strength that link the components together.
In subsequent experiments, we aimed to examine systematically the relationship between central cost and pattern stability, that is, between RT and relative phase variability.

5. Covariation of pattern stability and attentional cost

In Zanone et al.'s (2001) study, we aimed to obtain an finer description of the (co)variation between the stability of preferred bimanual patterns and central cost. We manipulated the oscillation frequency over a large range about that adopted spontaneously. The lower stability of the anti-phase pattern is not only evident before the transition, but also when the system is forced away from its "preferred" frequency, that is, when oscillations are performed at frequencies that are different from those adopted spontaneously (Schmidt et al., 1993). This suggests that there exists a task-specific comfortable frequency, at which bimanual patterns are more stable and, assumedly, less "costly" than at higher or lower frequencies.. Using a similar dual-task paradigm as above, we aimed to determine how much pattern stability and RT cost function match. In the case of a tight relationship, it may be assumed that both variables share a common, high-order dynamics. For both in-phase and anti-phase patterns, we expected loss of stability around the preferred frequency, manifested by an increase of relative phase variability. Central cost should be minimized at the preferred frequency of oscillation and a concomitant increase in this cost should be observed when the bimanual system moves away from the preferred frequency. In sum, we should observe a quadratic relationship of both pattern stability and reaction time, as a function of oscillation frequency.

The same experimental setup as in the previous study (Temprado et al., 1999; see above) was used, except for two buttons placed under both participants' feet (instead of fingers) that were used to record the RT response. Participants were dispatched between two groups. One group performed the in-phase pattern and the other performed the anti-phase pattern. In dual-task conditions, participants performed ten experimental conditions, with two priority levels and five different (randomized) frequencies: 0.5, 0.75, 1.0, 1.5, and 2.0 Hz. We carried out a polynomial regression for each pattern (in-phase and anti-phase) and for each priority condition on the mean SD of relative phase and on the mean RT obtained at each level of frequency. This analysis aimed to determine whether the modification of the oscillation frequency altered in a similar fashion the stability of the bimanual pattern (i.e., SD of relative phase) and the attentional cost (i.e., mean RT) associated with performing and/or stabilizing the coordination. Figure 4 indicates that all our variables exhibited a comparable evolution as function of oscillating frequency. Mean SD and RT for the in-phase and anti-phase patterns revealed a significant quadratic trend (cf. Figures 4A to D, respectively).

To further establish the similarity between these quadratic trends, we performed a cross-correlation between the results obtained for SD of relative phase and RT, as a function of frequency, for all experimental conditions. For the in-phase pattern, the largest correlation was found for a lag 0 analysis (cf. Figs 4A and B). For the anti-phase pattern, the largest correlation coefficient corresponded also to a lag 0 (cf. Figures 4C and D). Such results indicate that the curves overlap best as presented. In other words, the minimum of the quadratic trends for both patterns and variables is the same for the priority and non-priority conditions. For all conditions and patterns there is a curvilinear covariation of relative phase stability and attentional cost, as a function of oscillation frequency. These results show that there is an optimal frequency for which a best compromise between the fluctuations allowed to the pattern and the central cost incurred to sustain it is attained. They also show that the preferred frequency is more difficult to identify for the in-phase pattern, due to a more tenuous increase in variability provoked by frequency manipulation. At any rate, the tendency for the in-phase pattern follows

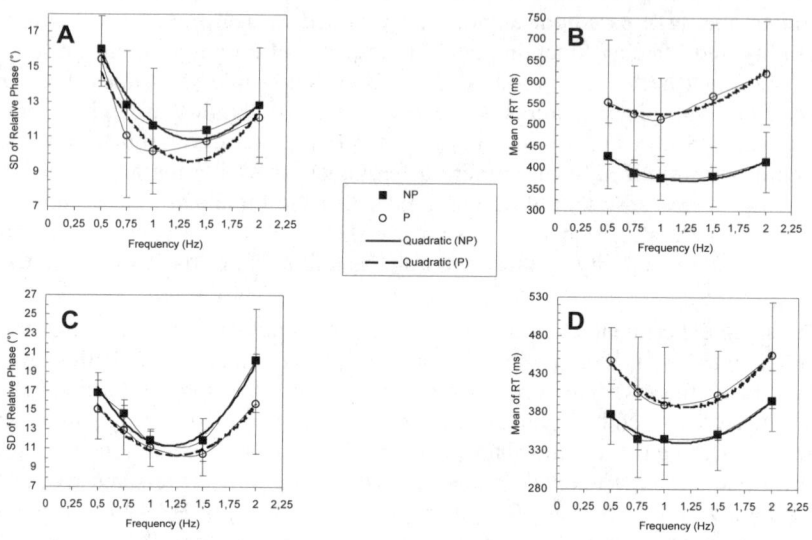

Figure 4. Average performance in the two tasks (bimanual coordination and RT) for the two dual-task conditions (Condition DU: shared attention; Condition DUBI: attentional priority given to the bimanual task) as a function of the level of frequency. Vertical bars ± 1 between-subjects standard deviation. **A**: The standard deviation of mean relative phase for the in-phase pattern. **B**: The mean RT, as an index of the attentional load associated to the in-phase pattern. **C**: The standard deviation of mean relative phase for the anti-phase pattern. **D**: The mean RT, as an index of the attentional load associated to the anti-phase pattern.

more a flat U-shaped and asymmetric evolution, rather than a sharp clear-cut V-shape trend. This suggests that the stability of the in-phase pattern is more affected by low than high frequencies. In contrast, for the anti-phase pattern, the curve is symmetric, although loss of stability seems to be faster for high than low frequencies. These findings are in keeping with the effects of spontaneous transitions induced by increasing oscillation frequency shown in the now classical experiment a la Kelso (Kelso, 1984, Kelso et al., 1986). In those studies, as in ours, such an effect was almost nil for the in-phase pattern.

Finally, we have also shown that the central cost incurred by the CNS to perform preferred bimanual patterns depends on their intrinsic stability. Indeed, RT and the SD of relative phase revealed U-shaped evolution as a function of frequency in the dual-task conditions, for both patterns of coordination. These quadratic trends express the covariation of pattern stability (SD of relative phase) and attentional cost (RT) about minimal values. Finer analyses showed that such minima coincide closely, so that the lowest variability and cost are found at the same frequency. These findings are in keeping with those pertaining to locomotion for metabolic cost (Hoyt and Taylor, 1981; Diedrich and Warren, 1995). Accordingly, performing a bimanual coordination pattern at the preferred frequency entailed a minimal attentional cost and a maximal stability of the coordination between the hands. Still, as raw an index RT may be, our paradigm pins down the collective activity of the CNS through a cost function, which depends, here, on oscillation frequency, thereby unveiling the dynamics that underlie both the cost function and the behavioral stability seen at the level of the coordination patterns.

6. Short-term effects of attention focus on behavioral dynamics and central cost.

In another study, we (Monno et al., 2000) aimed to establish whether the central cost incurred by the CNS, indexed by RT measures, covaries with the stability of the coordination patterns performed in the vicinity of (before and after) the phase transition from anti-phase to in-phase. We investigated the effects of focus of attention on bimanual coordination dynamics, especially on phase transitions between preferred stable coordination patterns induced by frequency manipulations. Following the paradigm pioneered by Kelso (1984), we hypothesized that the increase in pattern stability resulting from an attentional focus on the coordination pattern should significantly delay the transition process. We also aimed to determine whether relative phase variability and central cost pertain to the same high-order dynamics around phase transition. One expects an increase in RT before transition and a decrease in RT after the switching.

Analyses carried out on the percentage of phase transitions and relative phase variability showed that phase transitions occurred for both priority conditions, but were significantly delayed by increased attention, because of the enhancement in stability due to attentional focus. Indeed, the number of phase transitions significantly decreased for the priority condition. Analysis performed on the SD of relative phase of the anti-phase pattern (i.e., excluding the variability of the in-phase pattern after the transition) showed that focusing attention on the bimanual task significantly decreased SD of the relative phase irrespective of the performed frequency (Figure 5A). A comparison of the variability of the performed patterns before and after the transition (anti-phase and in-phase, respectively) indicated that, for both priority conditions, the variability of the anti-phase pattern first increased with increasing frequency before the phase transition and decreased thereafter. Analysis of mean time before transition also showed that attentional focus on the bimanual coordination significantly delayed the occurrence of the transition with respect to the non-priority condition. Analyses carried out on RT revealed that focusing attention on the bimanual coordination significantly increased RT, irrespective of the performed frequency (Figure 5B). Moreover, for both priority conditions, results showed that RT decreased after the transition to the in-phase pattern.

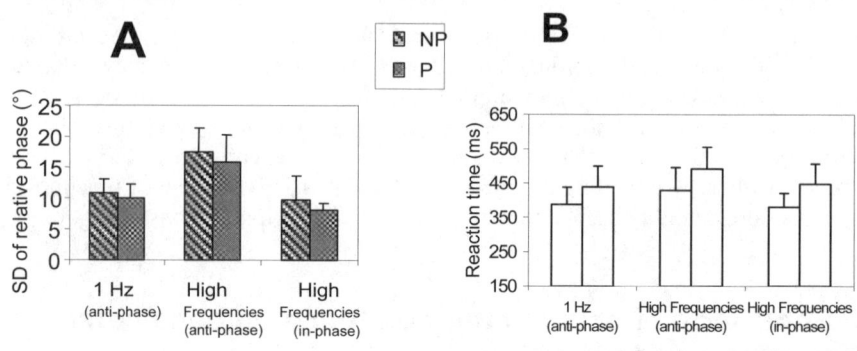

Figure 5. A: Standard deviation (SD) of relative phase of anti-phase and in-phase patterns in the non priority (NP) and priority (P) conditions for low and high frequencies. **B**: Mean RT for the anti-phase and in-phase patterns in the non priority (NP) and priority (P) conditions for low and high frequencies

Overall, the above results indicate that pattern stability and the central cost expended by the CNS to sustain preferred bimanual patterns do covary around phase transition. This suggests that the coordination dynamics influences both levels of overt coordinated behavior and of CNS processing activity. That the dynamics somehow paralleled does not imply that the central cost is the proximal cause of transitions between patterns. Rather, phase transitions are a consequence of loss of stability of the anti-phase pattern. Actually, the question arises of whether central cost and pattern stability reflect the same underlying dynamics or whether one of the dynamics is constrained by the other. A glimpse into the issue is afforded by considering separately the trials in which transitions did occur or not. For both priority conditions, analyses comparing the variability of relative phase for the trials with and without transitions revealed no significant differences. These results suggest that, in both conditions, the occurrence of a phase transition cannot be predicted univocally by the relative phase variability profile of the trial. However, in the priority condition, a significant difference in RT was observed between the trials with and without transitions: RT was significantly longer for the trials with transition. This indicates that two different mechanisms might induce phase transitions in the non-priority and priority conditions. In the shared attention condition, phase transitions occur as a result of loss of stability of the anti-phase pattern. In the focused attention condition, pattern stability is so strong that transitions may be due to too hefty a central load needed to sustain the anti-phase pattern at higher frequencies. Thus, the central cost associated with attentional focus might act as a trigger for phase transitions in bimanual coordination.

7. Co-Evolution of pattern stability and central cost with learning

In all the experiments reported above, attention has been shown to display the characteristics of *behavioral information*, influencing temporarily the stability of preferred patterns stability. Then, the present approach also provides an operational means to study long-term effects of practice on spontaneous behavioral coordination dynamics and central cost. In a dynamic pattern framework, learning is considered the process by which behavioral information becomes memorized behavioral information (Schöner and Kelso, 1988a, 1998b; Schöner et al., 1992). Once learning is achieved, the memorized pattern constitutes an attractor of the behavioral pattern dynamics. Zanone and Kelso (1992) followed this line of reasoning to study how relative phase evolved according to specific behavioral requirement of relative phase (i.e., 90°) that did not correspond to the spontaneous dynamics. However, to our knowledge, the evolution of preferred pattern stability and spontaneous dynamics with learning has never been studied before.

It has long been argued that the acquisition of complex skill over practice is, at least partially, accompanied with a change in the amount of information that can be processed simultaneously at a central level. However, what kind of modifications in the CNS information processing activity mediates overt changes in attentional demands with learning inter-limb coordination is not well established. In a recent study (Temprado et al., 2002), we aimed to investigate the effects of learning on bimanual spontaneous coordination dynamics and attentional demands. Participants were asked to execute a cyclic *anti-phase* coordination pattern (180° of relative phase), while performing a *reaction time task* (RT) by depressing simultaneously two trigger buttons with their feet. A pre-test permitted to assess the spontaneous coordination dynamics and to determine each individual critical transition frequency. In the acquisition phase, participants were trained to maintain the anti-phase coordination pattern at the critical oscillation frequency. After thirty minutes and seven days, a post-test and a retention test were performed with the same procedure as in the pre-test.

Analysis carried out on the percentage of phase transitions showed that the number of transitions decreased in the post-test and the retention test, as a result of learning. Analysis of the first exit time showed the same trend: Time before transition increased for the post-test and the retention test, as a result of learning. Moreover, after the acquisition phase (post-test and retention test), subjects were able to maintain the anti-phase pattern at a higher frequency than in the pre-test (Figures 6A,B).

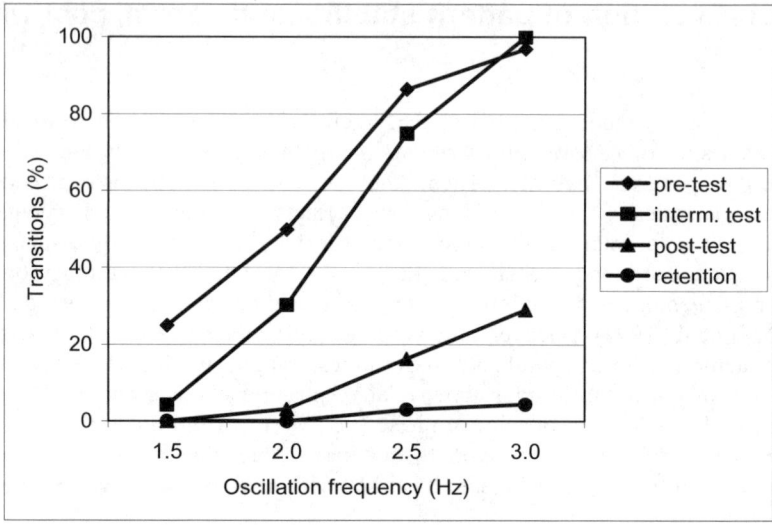

Figure 6. A: Percentage of trials with transition in the pre-test, the post-test and the retention test, as a function of frequency. **B**: Time (s) before transition occurred in the pre-test, the post-test and the retention test, as a function of frequency.

Analysis of SD of relative phase showed that the anti-phase pattern stability increased for the post-test and the retention test, for all levels of frequency but 3.0 and 3.5 Hz (Figure 7A). Surprisingly, although relative phase variability was equivalent to the pre-test at this frequency, a decrease in the number of phase transitions was still observed. Analysis of Reaction Time (RT) showed the same trend: RT decreased at the end of the acquisition phase for all frequency levels but 3.0 Hz and 3.5 Hz.

However, this effect was only observed for the retention test (Figure 7B), suggesting the existence of consolidation processes that occurs between sessions and overnight to enhance the effects of practice (Brashers-Krug et al., 1996; Shadmehr and Brasher-Krug, 1997; Shadmehr and Holcomb, 1997). Moreover, although relative phase variability continued to increase at high frequencies, RT roughly stabilized. This suggests that fluctuations do not predict univocally phase transition. Rather, attentional demands associated to performing the anti-phase pattern may act on the (non) occurrence of phase transitions, at least at high frequency levels (for converging evidence, see Monno et al., 2000).

The above results show that long-lasting changes in spontaneous coordination patterns may be induced by learning. Practice of the anti-phase pattern at a critical frequency results in an increase of relative phase stability at low and intermediate frequencies; it also results in a decrease of the number of phase transitions and an increase of time before transition for all the frequency levels. This study also indicates that pattern stability and the central cost expended by the CNS to sustain preferred bimanual patterns covary with learning. Such a decrease of attentional

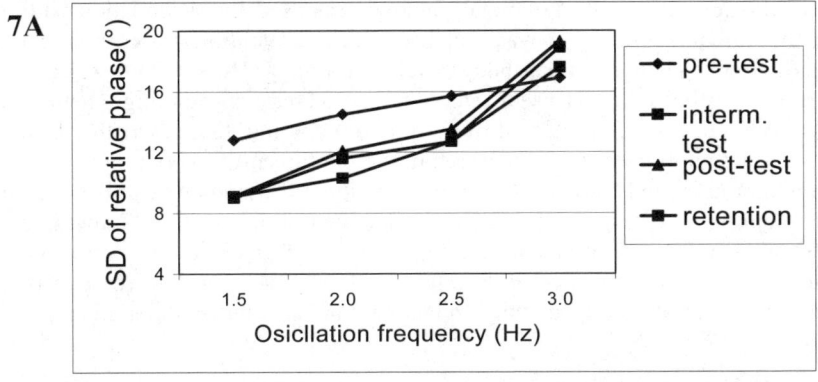

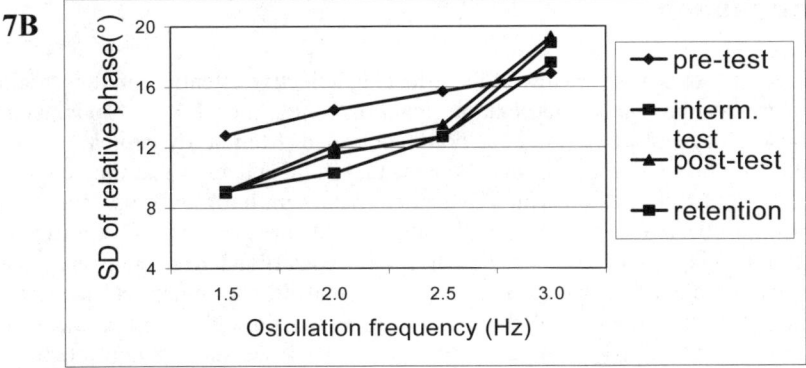

Figure 7. A: Standard deviation (SD) of the mean relative phase of anti-phase pattern for the pre-test, the post-test and the retention test, as a function of frequency. **B**: Mean RT of anti-phase pattern for the pre-test, the post-test and the retention test, as a function of frequency.

cost can be considered a somewhat trivial result for cognitive psychologists and neuroscientists. However, underlying mechanisms of this decrease remain largely unknown, particularly in perceptivo-motor and inter-limb coordinations. This study and those reported above suggest that central cost incurred by the CNS to maintain the anti-phase pattern mainly depend on its stability. These results corroborate, using a measure of attentional processing demands, those obtained in recent neurophysiological studies which have described neural correlates of the stability and change of behavioral coordination using high density SQUID and EEG arrays (Kelso et al., 1992; Mayville et al., 1999, 2001) as well as fMRI (Fuchs et al., 2000). In a recent study, Jantzen et al. (2001) concluded that training altered the way in which the brain performs the task. Specifically, they showed that desynchronization of oscillatory activity of the neural ensembles of the cerebral cortex is an appropriate measure of learning. Jantzen et al.'s study (2001) also led to the hypothesis that the post-training CNS activity differences reflect a decrease in the associated attentional demands imposed by syncopation pattern. Thus, the authors' conclusion was that learning and attention are co-implicative and may involve changes in cell-body/synaptic coupling. However, in their study, Jantzen et al. (2001) did not use a retention test, thereby questioning whether the effects of practice were permanent over some time period, that is whether change in cortical activity observed following training continued to evolve over days. In Temprado et al.'s study, the reduction of attentional demands was only observed in the retention test, suggesting a delayed effect of training on the RT measure of attentional demands. Thus, an interesting issue for further studies would be to couple brain imaging and dual-task experiments to assess on-line (parallel) and delayed evolution of cortical activity and attentional demands with learning.

8. Conclusion

This paper presents an overview of the effects of selective attention on the stability of preferred bimanual coordination patterns and, hence, on spontaneous coordination dynamics, a topic seldom addressed from a dynamical systems approach. An originality of the studies presented in this paper is that they combine a classical dual-task method with a dynamical paradigm in order to investigate the relationships between attention and the control or learning of coordinated behavior. This procedure establishes a correspondence between two conceptualizations of coordination, until now isolated: The dynamical landscape, defined by the coordination dynamics and assessed through stability measures, and the information processing activity of the CNS, assessed through attentional cost measures (i.e., Reaction Time). We assumed a relationship between dynamics by virtue of a link between dynamical models (HKB; S and K) and the energetic metaphor of attentional resources. We hypothesized that the assessment of the central cost through the dual-task paradigm affords a (fairly) direct and reliable evaluation of the activity devoted by the CNS to maintain the coordination pattern. Overall, our results showed that bimanual coordination and the processing activity ensured by the CNS to maintain the stability of such coordination follow the same dynamics. These results provide new insights regarding the relationships between attentional demands, brain processes and bimanual coordination pattern dynamics. They show that attention is *in* action; that is it resides within the self-organizing dynamics of behavior, forming an indissociable part of such dynamics. Then, the observed co-variation of pattern stability and attentional demands seems to reflect a general principle that link coordination dynamics and central processing activity. This principle may capture the means by which the CNS stabilizes or destabilizes coordinated behavior in a task-specific, informationally meaningful fashion.

Acknowledgments

We sincerely thank A. Chardenon, T. Coyle and M. Lecore, M. Laurent, A. Monno and P.G. Zanone for their contribution to this research program.

References

Amazeen EL, Amazeen PG, Treffner PJ, Turvey MT (1997) Attention and handedness in bimanual coordination dynamics. J Exp Psychol Human 23, 1552-1560

Brashers-Krug T, Shadmehr R, Bizzi E (1996) Consolidation in human motor memory. Nature 382, 252-255

Carson RG, Goodman D, Kelso JAS, Elliott D (1994) Intentional switching between patterns of interlimb coordination. J Hum Movement Stud 27, 201-218

Carson RG, Byblow WD, Abernethy B, Summers JJ (1996) The contribution of inherent and incidental constraints to intentional switching between patterns of bimanual coordination. Hum Movement Sci 15, 565-589

Diedrich FJ, Warren WH (1995) Why change gaits? Dynamics of the walk-run transition. J Exp Psychol Human 21, 183-202

Fuchs A, Mayville JM, Cheyne D, Weinberg H, Deecke L, Kelso JAS (2000) Spatiotemporal analysis of neuromagnetic events underlying the emergence of coordinative instabilities. NeuroImage 12, 71-84

Haken H, Kelso JAS, Bunz H (1985) A theoretical model of phase transition in human movements. Biol Cybern 51, 347-356

Hoyt DF, Taylor CR (1981) Gait and energetics of locomotion in horses. Nature 292, 239-240

Jantzen KJ, Fuchs A, Mayville JM, Deecke L, Kelso JAS (2001) Neuromagnetic activity in alpha and beta bands reflect learning-induced increases in coordination stability. Electroen Clin Neuro 112, 1685-1697

Kahneman D (1973) Attention and effort. Prenctice-Hall, Englewood Cliffs, NJ

Kelso JAS (1984) Phase transitions and critical behavior in human bimanual coordination. Am J Phys, Reg I 15, R1000-1004.

Kelso JAS (1995) Dynamic Patterns. The Self-organization of Brain and Behavior. The MIT Press, Cambridge, MA

Kelso JAS, Bressler SL, Buchanan S, DeGuzman GC, Ding M, Fuchs A., Holroyd T (1992) A phase transition in human brain and behavior. Phys Lett A 196, 134-154

Kelso JAS, DelColle JD, Schoner G (1990) Action-perception as a pattern formation process. In: Jeannerod M (ed.) Attention and Performance XIII. Lawrence Erlbaum, Hillsdale, NJ, 139-169

Kelso JAS, Scholz JP, Schöner G (1986) Nonequilibrium phase transitions in coordinated biological motion: critical fluctuations. Phys Lett A 118, 279-284

Lee TD, Blandin Y, Proteau L (1996) Effects of task instructions and oscillation frequency on bimanual coordination. Psychol Res 59, 100-106

Mayville JM, Bressler SL, Fuchs A, Kelso JAS (1999) Spatiotemporal reorganization of electrical activity in the human brain associated with a timing transition in rhythmic auditory-motor coordination. Exp Brain Res 127, 371-381

Mayville JM, Fuchs A, Ding M, Cheyne D, Deecke L, Kelso JAS (2001) Event-related changes in neuromagnetic activity associated with syncopation and synchronization timing tasks. Hum Brain Mapp 14, 65-80

Monno A, Chardenon A, Temprado JJ, Zanone PG, Laurent M (2000) Effects of attention on phase transitions between bimanual coordination patterns: A behavioral and cost analysis in humans. Neurosc Lett 283, 93-96

Navon D (1990) Exploring two methods for estimating performance tradeoff. Bull Psychon Soc 28, 155-157

Parasuraman R (1998) The attentive brain. MIT Press, Cambridge

Sanders AF (1998) Elements of human performance. Reaction processes and attention in human skills. MIT Press, Lea, NJ

Shadmehr R, Brashers-Krug T (1997) Functionnal stages in the formation of human long-term memory. J Neurosc 17, 409-419

Shadmehr R, Holcomb HH (1997) Neural correlates of motor memory consolidation. Science 277, 821-825

Schmidt RC, Shaw BK, Turvey MT (1993) Coupling dynamics in interlimb coordination. J Exp Psychol Human 19, 397-415

Scholz JP, Kelso JAS (1990) Intentional switching between patterns of bimanual coordination depends on the intrinsic dynamics of the patterns. J Motor Behav 22, 98-124

Schöner G, Haken H, Kelso JAS (1986) A stochastic theory of phase transitions in human hand movement. Biol Cybern 53, 442-452

Schöner G, Kelso JAS (1988a) A dynamic theory of behavioral change. J Theor Biol 135, 501-524

Schöner G, Kelso JAS (1988b) A synergetic theory of environmentally-specified and learned patterns of movement coordination. Biol Cybern 58, 71-80

Schöner G, Zanone PG, Kelso JAS (1992) Learning as change of coordination dynamics: Theory and experiment. J Motor Behav 24, 29-48

Swinnen SP, Jardin K, Meulenbroek R (1996) Between-limb asynchronies during bimanual coordination: Effects of manual dominance and attentional cueing. Neuropsychology 34, 1203-1213

Temprado JJ, Monno A, Zanone PG, Kelso JAS (2002) Attentional demands reflect learning-induced alterations of bimanual coordination dynamics. Eur J Neurosci 16, 1390-1394

Temprado JJ, Zanone PG, Monno A, Laurent M (1999) Attentional load associated with performing and stabilizing preferred bimanual patterns. J Exp Psychol Human 25, 1579-1594

Tsang PS, Shaner TL, Vidulich MA (1995) Resource scarcity and outcome conflict in time-sharing performance. Percept Psychophys 57, 365-378

Tsang PS, Velazquez VL, Vidulich MA (1996) Viability of resource theories in explaining time-sharing performance. Acta Psychol 91, 175-206

Van der Molen MW (1996) Energetics and the reaction process: Running threads through experimental psychology. In: Neumann O, Sanders F (eds.) Handbook of perception and action, Vol 3. Academic Press, London, 229-276

Wickens CD (1984) Processing resources in attention. In: Parasuraman R, Davies DR (eds.) Varieties of attention. Academic Press, San Diego, 63-98

Wuyts IJ, Summers JJ, Carson RG, Byblow WD, Semjen A (1996) Attention as mediating variable in the dynamics of bimanual coordination. Hum Movement Sci 15, 877-897

Zanone PG, Kelso JAS (1992) Evolution of behavioral attractors with learning: Non equilibrium phase transitions. J Exp Psychol Human 18, 403-421

Zanone PG, Kelso JAS (1997) The dynamics of learning and transfer: Collective and component levels. J Exp Psychol Human 23, 1454-1480

Zanone PG, Monno A, Temprado JJ, Laurent M (2001) Shared dynamics of attentional cost and pattern stability in the control of bimanual coordination. Hum Movement Sci 20, 765-789

Intention in Bimanual Coordination Performance and Learning

Timothy D. Lee

Department of Kinesiology, McMaster University, Hamilton, Ontario, Canada
scapps@mcmaster.ca

George Miller used to say to graduate students, "I can foul up any one of your experiments by an act of will" (Miller, 1986 p 221). Although Miller was making an editorial comment on the behaviorist tradition, an important lesson for all was that even the most clever experimental design can be compromised by the intentions of the research participant (deliberate or not).
Experiments in motor control and learning are no less susceptible to the intentions of the individual, and in some cases, can reveal rather important findings when strong "effects" in motor control are enhanced, neutralized, or even reversed by the intentions of the participant. Experimental manipulations that attempt to influence a participant's "will" through instructions are often expressed in terms of a particular movement goal, strategy, or attentional focus.
I attempt to address the following questions in this chapter: 1) what are some of the influences of intention and attentional focus on motor control and learning? and 2) of what importance are these influences to our understanding of motor control and learning? In concert with the theme of this book, coordination research will be the window into these answers. The first part of the chapter focuses on the influence of intention in the performance of stable coordination patterns. Learning a new coordination pattern is the focus later in the chapter.

1. Performance of stable coordination patterns

Two bimanual coordination patterns dominate behavior when the upper limbs are coordinated in relative time (excellent reviews of this literature are available and might help to fill in some background issues not covered here – see Kelso, 1995; Swinnen, 2002; Turvey, 1990). In one pattern ("in-phase" coordination), simultaneous activation of homologous muscles (e.g., simultaneous tapping of the index fingers of each hand) can be performed very accurately and consistently. In the other pattern ("anti-phase" coordination), the alternate activation of homologous muscles (e.g., alternate finger tapping) also results in very accurate and stable coordination patterns (Kelso, 1984). Although the performance characteristics of these two patterns appear similar under comfortable (or preferred) conditions, a difference is revealed when the movement system is challenged. For example, the characteristics of these patterns differ when an individual is asked to move the upper limbs at a movement frequency that is faster than what might be considered to be comfortable by an individual. At increased movement frequencies the in-phase pattern remains very accurate and stable, but the anti-phase pattern does not. The influence of metronome frequency, as an event that directly changes the characteristics of these coordination patterns, is considered by many to be non-specific (or non-intentional) in terms of its influence (a "control parameter" in dynamic pattern theory terms). However, as will be seen, the expression of metronome frequency in the behavior of the individual is very much influenced by intentions. In the next three sections I present three different lines of evidence which suggests that the accuracy and stability of both in-phase and anti-phase patterns is subject to the influences of the intentions of the individual.

1.1 Instructions and pattern transitions

As mentioned previously, the characteristics of the anti-phase pattern (but not the in-phase pattern) change when the movement system is challenged, such as in increased metronome frequency conditions. However, research has also shown that the characteristics of this change in the anti-phase pattern interact with the intentions of the individual (i.e., under different instructions). In some experiments, individuals have been instructed to emphasize keeping pace with the frequency of the perceptual stimulus (metronome). If they begin to "lose" the originally intended coordination pattern these subjects have been told to behave in a particular way – they have been instructed to not intentionally intervene in order to remain in that pattern, and to perform the pattern that feels the most comfortable at that speed. Under such instructions, the typical finding is that the anti-phase pattern becomes so unstable at a critical movement frequency that the movement system finds a coordination solution by restabilizing itself as an in-phase pattern (e.g., Kelso, 1984). This characteristic, for the system to make an unintentional, non-linear transition in relative phase, is typical of performance given this behavioral instruction.

It is also important to note that non-linear transitions in relative phase can be unintentional (as above) or intentional. In the latter case, experiments have been conducted in which participants are asked to make a purposeful transition from anti-phase to in-phase (or vice versa). In such volitional transition cases, the switch can be accomplished smoothly, following a destabilization of the original pattern (Scholz & Kelso, 1990; Byblow et al., 1999).

An interesting issue emerges however, when individuals are asked to intentionally intervene to stay in the original pattern when the stability of the coordination pattern begins to be lost (e.g., at increased movement frequencies). Under these instructions, the performance of the anti-phase pattern becomes less stable but does not undergo a pattern switch to in-phase. Rather, the fate of the anti-phase pattern is an increased instability rather than in terms of a return to stability following a non-linear transition (Lee et al., 1996; Wishart et al., 2000). The instruction to stay in the anti-phase pattern does not improve the stability of the anti-phase pattern, nor does it delay the critical frequency at which the disruption to the anti-phase pattern occurs (Smethurst & Carson, in press). Rather, the intention to attempt to stay in anti-phase results in a central tendency of relative phasing about 180°, with large deviations from this central tendency.

The argument here is that instructions influence the individual's intent to perform with a certain expression of motor behavior (i.e., anti-phase). The intentions of the individual do not override the inherent dynamics; nor do the dynamics determine the expression of motor behavior in the absence of the individual's intention. It is the interaction of intention and the inherent properties of coordination that seems to be most important (Kelso et al., 2001).

1.2 Handedness and directed attention

Although it might appear that the sensory and motor contributions of each limb in bimanual coordination are achieved by an equal "weighting", the fact remains that most humans have a predominant handedness, many more being right-hand than left-hand dominant. The contribution of this dominance in bimanual coordination is illustrated nicely in a number of examples provided by Peters (1994) and in the work of Carson, Byblow and colleagues. Experiments in which phase transitions occur have shown that the breakdown of the anti-phase pattern is almost always initiated by the non-dominant limb (e.g., Carson et al., 1996). The dominant limb is more apt to remain in its initially prepared timing mode than the non-dominant limb.

The influence of the dominant limb in the performance of bimanual patterns can also be mediated be intention, as illustrated well in an experiment by Amazeen et al. (1997). Groups of self-declared right-hand and left-hand dominant individuals performed in-phase coordination of two hand-held pendula. On different trials the subjects directed their attention to either their right or left limb, with augmented goal-related information about the spatial trajectories being provided for the limb to which attention was being directed. The resulting interaction of handedness and directed attention for the stability of the in-phase pattern is illustrated in Figure 1. For the right-hand dominant individuals, attention directed to the right (dominant) limb facilitated the coordinated performance of *both* limbs more than when attention was directed to the left limb. Similarly, for the left-hand dominant subjects, attention directed to the left (dominant) limb stabilized performance more than when attention was directed to the non-dominant limb. The contribution of handedness was that the normally accurate and stable in-phase pattern appeared either to be *enhanced* when attention was directed to the dominant limb or *disrupted* when attention was directed to the non-dominant limb. Once again it is seen that the behavioral expression of the inherent properties of the motor system, here in terms of hand dominance, can be mediated by the intentions of the individual. The intentions do not *change* the individual – a right-hand dominant person does become left-handed simply by intention. Rather, the interaction of intention and these inherent properties combine to determine performance.

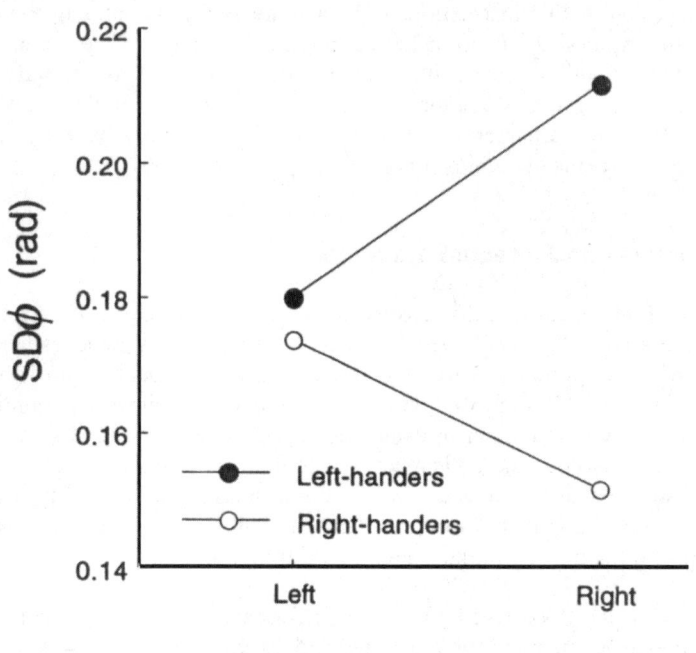

Figure 1. Effects of direction of attention on the stability of in-phase coordination for left- and right-hand dominant individuals (reprinted from Amazeen et al. 1997).

1.3 Directed attention in dual task performance

A more "classic" example of the effects of intention has been the research examining the effect of combining the performance of the in-phase and anti-phase coordination patterns with a secondary load task (Monno et al., 2000; Temprado, 2003; Temprado et al., 1999; Temprado et al., 2001; Zanone et al., 2001). For example, in the Temprado et al. (1999) study, trials of in-phase and anti-phase joystick movement patterns were sometimes combined with a task in which an auditory probe was responded to as rapidly as possible by pressing the trigger button on the joystick (a secondary RT task). An interesting variant of this procedure was also introduced in which the individual was also asked, on some trials, to give attentional "priority" to either the coordination pattern or the RT task.

The main findings from the Temprado et al. (1999) study are illustrated in Figure 2. In control (single task) conditions (two leftmost bars in Figure 2) the in-phase and anti-phase patterns were performed with relatively equal stability, which is typical for performance when paced at preferred movement frequencies. Performing the RT task in conjunction with the coordination task had no effect on the stability of the in-phase pattern, regardless of whether an intentional preference was given to the movement task or to the RT task (filled bars in Figure 2). In contrast, the stability of the anti-phase pattern seemed to depend on how attentional priority had been directed. When directed toward the coordination task, performance of anti-phase was nearly as stable as it was when no secondary task was performed. However, when attention was directed toward the RT task, the instability of the anti-phase pattern increased dramatically.

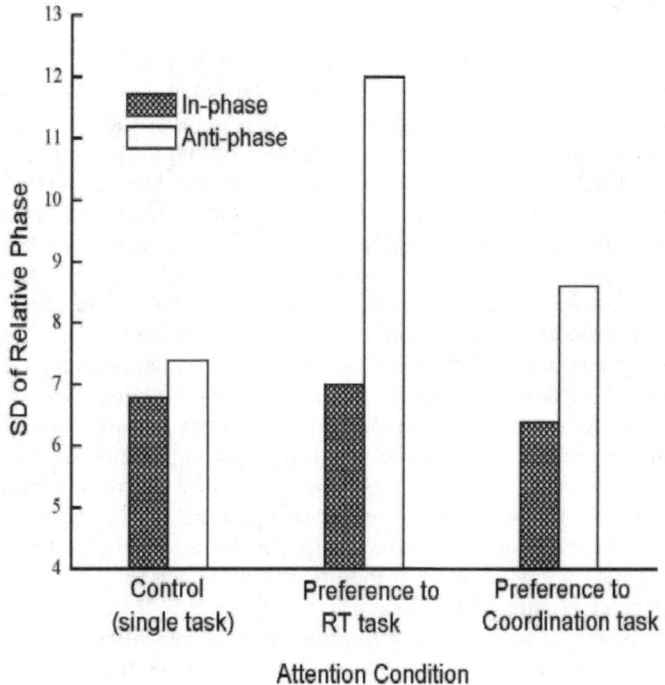

Figure 2. Effects of divided attention preference on stability of in-phase and anti-phase coordination (adapted from data presented in Temprado et al. 1999).

In some regards, this situation may be considered to be a triple task rather than the more typical dual task found in most experiments on attentional demands. The combined task in this experiment involved a movement timing task with the right hand, a movement timing task with the left hand, and a perceptual-motor RT task. As seen in the previous two sections, the behavioral expression of coordination task performance resulted from an interaction of the intentions of the individual and the inherent properties of the movement patterns. All three tasks could been combined with no detrimental effect when homologous muscles were used in the limb timing task (in-phase pattern). However, when the timing task involved the alternation of homologous muscles, the performance stability of the anti-phase pattern could only be achieved by assigning an attentional priority to these movements. As in the case with increased movement frequency effects (section 1.1), the fate of the stability of the anti-phase pattern interacted with the intentions of the individual.

1.4 Discussion of section 1: On the role of intention in coordination performance

A common issue pertaining to the roles of intention (or attention) in the above sections relates to something Bernstein discussed many years ago (reprinted in Bernstein, 1996). In the following quote, Bernstein discusses how an individual should appropriately channel conscious attention during the performance of a skill. "We know that consciousness always resides at the leading level for a given movement. All the processes taking place in background levels, all the automatisms and auxiliary corrections, proceeds outside its boundaries. Therefore, *to fix conscious attention to one of the background mechanisms* virtually always means *temporarily turning a corresponding background level into the leading one*, that is, introducing a disruptive switching (of levels)... attention should be focused on a level where consciousness resides and which takes the responsibility for movement success in its major and most important components. Therefore, one should concentrate on the *desire to solve a motor problem as accurately and expediently as possible*. This desire will lead to basic, meaningful corrections for the whole movement. For example, the attention of a person who has learned to ride a bicycle should not be fixed on his legs or arms but on the road in front of the bicycle; the attention of a tennis player should be directed at the ball, the top edge of the net, the movements of the opponent, but certainly not at his own legs or on the racket." (pages 201-203).

Bernstein appears to have been confirming what coaches often tell their athletes – focus on the goal and trust your movement skills. In experiments on coordination, it is the movement *pattern* that represents the problem to be solved, not the individual movements of each limb which are coordinated to define the pattern. In this sense, the pattern represents a more abstract, or in Bernstein's terms, a higher "level" problem to be solved.

Intention to stay in a coordination or to not intervene in a pattern switch represents a problem at the pattern level. Recall for example, from Smethurst and Carson study that intentions did not "suppress" the influence of the intrinsic dynamics because there remained the same attraction toward the in-phase pattern, nor did it "enhance" the intrinsic dynamics by somehow "strengthening" the anti-phase pattern. Perhaps instead what is happening is that the intention to stay in the initially prescribed pattern established a goal state against which information about the system (via internal and augmented feedback) can be compared and upon which action can be taken when error information exceeds a tolerance level. In contrast, the absence of an intention to stay in the pattern might establish only a weak goal state (sufficient to initiate the pattern) or to decrease the tolerance levels at which error information is acted upon. By this notion, intention is considered to reflect a desire or will to act upon behavioral information relative to the status of the coordination pattern as the problem, and to do so in a manner and intensity that is consistent with the conscious problem.

Although handedness plays a role in the determining the pattern, it could be the case that the natural abstractness of the in-phase pattern is disrupted when attention is directed to the non-dominant limb. The Temprado et al. (1999) experiment nicely illustrates what happens if attention is diverted away from the pattern as the problem to be solved. This evidence suggests that in-phase patterns can be maintained without full attention at the level of the pattern (although in-phase can be disrupted when attending to the non-dominant limb, as seen in Amazeen et al., 1997). However, the anti-phase pattern does appear to need more conscious attention directed at the level of the pattern in order to maintain stability, a finding that highlights the complex interaction of intention and the inherent properties of coordination dynamics.

2. Learning a new coordination pattern

As discussed previously, accurate and stable patterns of bimanual coordination (in-phase and anti-phase) can be performed at preferred tempos without practice. This observation has been seen many times in experiments involving young adults (see recent review by Swinnen, 2002). Stable coordination patterns are also seen in children by the age of one (e.g., Fagard, 1994), in healthy older adults (Greene & Williams, 1996; Serrien et al., 2000; Wishart et al., 2000), and in patients with neurological disorders such as Parkinson's disease (Almeida et al., 2002; Byblow et al., 2002; Verschueren et al., 1997). However, the relative timing of upper limbs other than in these two preferred coordinations is difficult, if not impossible to achieve without practice (Yamanishi et al., 1980). The most prominent difficulty in producing a coordination pattern other than in-phase or anti-phase is in "breaking away" from these preferred patterns – bimanual timing naturally tends to drift toward one or the other of these patterns. Therefore, learning a new coordination pattern, involves the processes of both acquiring information about what to do and how not to perform the preferred patterns.

2.1 Intention during learning as influenced by the task demands

Accurate and stable patterns of coordination other than in-phase and anti-phase can be achieved readily if an individual is provided with sufficient practice and augmented feedback information (e.g., Zanone & Kelso, 1992, 1997). In such a case the concept of "intention" as it relates to the expression of a bimanual coordination pattern can take on different meanings. For instance, individuals in the experiment by Zanone and Kelso (1992) practiced a 90° pattern of relative phasing (equidistant "between" in-phase and anti-phase) by trying to match the finger movements on each hand to two discrete, visual metronomes that were offset in relative time by 1/4 of a cycle. By this method, it could be argued that the intentions of the individual were to learn a perceptual-motor, spatial-temporal coordination between the visual metronomes and the finger oscillations. Since the visual metronomes were offset by 90°, a successful coupling of each finger oscillation with its metronome would achieve the "intention" of a 90° relative phase pattern.

A different approach to learning a new coordination pattern has been used in other experiments (Fontaine, Lee, & Swinnen, 1997; Lee, Swinnen, & Verschueren, 1995; Smethurst & Carson, 2001; Swinnen et al., 1998). In these experiments the learner was typically given more explicit information about what to do to perform a new coordination pattern than in the Zanone and Kelso studies. For example, the learner may be explained the relationship between the limbs in performing in-phase and anti-phase patterns, and that a 90° pattern, for example, would require a relative timing that was "between" these two preferred, natural patterns. In some cases, individuals in these experiments were provided with augmented feedback in the form of a Lissajous Figure (a real-time displacement-displacement plot of the two limbs), presented either concurrently with the learner's movements or as terminal feedback (e.g., Swinnen et al., 1997). Thus, in this approach it could be argued that the intentions of the individual are to learn a sensory-motor relationship between the two limbs, rather than between each limb and a visual signal.

A third experimental approach has been used in which a dynamic, continuous visual stimulus, presented as a 90° Lissajous Figure (a circle), was presented during the practice trial and the individual was asked to try to match their augmented feedback trace (as a Lissajous Figure) to the Lissajous Figure stimulus. Essentially, such a procedure makes bimanual coordination to be a type of pursuit tracking task. In experiments such as these (Kurtz & Lee, in preparation; Wenderoth & Bock, 2001), one could argue that the intentions of the individual are much more perceptual in purpose – to match the feedback signal to the stimulus signal.

The importance of these methodological differences and potentially, the differences in the intentions of the learner that result from these methodologies, are important when trying to account for various differences in the learning profiles of new coordination patterns. In the research of Zanone and Kelso (1992 1997) for example, the learning of a new coordination pattern resulted in the destabilization of one of the inherently stable patterns (anti-phase). This destabilization was found to be only a temporary effect in other research however (Fontaine et al., 1997; Lee et al., 1995; Smethurst & Carson, 2001). We suspect that the methodologies and intentions of the learner that were influenced under one set of methods could be operating in quite a different way than under another set of methods. For instance, anti-phase performance in one methodology (Zanone & Kelso) is much more perceptually demanding than in the other methodology. Perhaps what has been learned about the new pattern in the Zanone and Kelso task competes more with the anti-phase pattern than in the methods where the learner is focusing more on the sensory-motor relationship between the limbs.

Another example is the difference in learning profiles exhibited by younger and older adults. Younger adults have been found to be proficient in learning a new coordination pattern under a variety of experimental conditions, including those providing sparse augmented information. In contrast, healthy older adults have considerable difficulty in learning a new coordination pattern unless considerable explicit augmented information is provided about the task and regarding the ongoing performance (e.g., Swinnen et al., 1998; Wishart et al., 2002). Again, these differences in learning style speak to potential differences in what the learner was intending to do given these task constraints.

2.2 Intention during learning as influenced by instructions

One might suppose that the methodological differences discussed in the learning research in the previous section may have created a different focus of attention amongst learners, possibly accounting for some of the differences found in the results of this research. For example, in an informationally rich environment, such as "tracking" task (Wenderoth & Bock, 2001), it could be reasoned that the learner's attention is focused on the perceptual display – information about the limb movements being projected and interpreted based on the mismatch between the visual traces of the stimulus and the feedback. In the absence of such information (e.g., Lee et al., 1995), knowledge about how performance is progressing can only be obtained by monitoring the visual and kinesthetic feedback arising from the motions of the limbs, and then acting on this information.

The potential importance of focus of attention effects in studies of bimanual coordination might be most profound when considering the role of the inherently stable patterns during learning. Recall the earlier discussion that, in learning a new coordination pattern, the biggest initial problem is in breaking away from the tendency to drift into an in-phase or anti-phase pattern. Before discussing the related research on this issue it is important to note that focus of attention has only recently become an emergent theme in the motor learning literature, despite the fact that Bernstein and others (e.g., James, 1896) had commented on this very issue many years ago. Largely due to the work by Wulf and her colleagues (reviewed in Wulf & Prinz, 2001), these studies have typically compared instructions that direct the learner's attention toward the sensory and/or motor aspects of the limbs involved in the task (an "internal" locus of attention) versus instructions that direct the learner's attention to a component of the task that is the object of the limb movements (an "external" locus of attention). For example, in a study examining the learning of a golf pitch shot by novices, one group was instructed to think about swinging the arms during performance (an "internal" focus) and compared to another group that was instructed to think about the pendulum motion of the golf club during the shot (an "external" focus). Studies comparing similar instruction groups using tasks such as tennis, platform balancing, and ski simulation have produced the same pattern of results: an external locus of attention produces superior performance in practice, superior retention, and sometimes superior transfer compared to an internal locus of attention (Wulf & Prinz, 2001).

Based on research conducted by Wulf and others it might be expected that learning a new bimanual coordination pattern would be facilitated by instructing the individual "away" from an internal focus of attention. Indeed, some bimanual coordination research seems to support this expectation. In an experiment by Hodges and Lee (1999), four groups of learners were each provided differing sets of instructions regarding the acquisition of a 90° relative phase pattern. All individuals in the study were told to try to produce a circle on the computer monitor (the correct Lissajous Figure for 90° relative timing) while keeping pace with an auditory metronome (at 1 Hz). Subjects in the "No Instructions" and the "Secondary task" groups received no further information about how to do the task, with the latter group given the additional responsibility of performing a subtraction task while attempting to perform the new coordination pattern. Subjects in both the "General" and "Specific Instructions" groups were provided with additional information about how to perform the 90° pattern. The "General Instructions" group was told that an appropriate strategy for acquiring the new coordination pattern would be to try to lag one limb by 1/4 of a movement cycle relative to the other limb. In addition to this general instruction, the "Specific Instructions" group were also given a detailed pictorial representation of the spatial relationship between the limbs at each movement reversal point for both limbs. These instructions were presented to the individuals in these two groups twice on each of two days of practice. The rationale was to attempt to direct the learner to think about the spatial-temporal relationship between the limbs during performance (i.e., an internal focus of attention).

The mean error results of the Hodges and Lee (1999) experiment, illustrated in Figure 3, revealed that the effect of the specific and general instructions was to bias the performance of the new coordination pattern toward 180° – in the direction of the anti-phase pattern. This bias was particularly strong at the beginning of practice and was not entirely overcome even after two days of practice. Performing a secondary task during practice drove the bias in performance toward 0° – in the direction of the in-phase pattern. However, this bias was overcome by the end of practice on Day 1. Performance on the new task was biased the least and overcome the fastest by the No Instruction group. Interestingly, although all groups performed the retention test well, a transfer test to a previously unpracticed new coordination pattern (45°) was most poorly performed by the general and specific instruction groups. A similar pattern of results for transfer was also seen in the performance stability measure. If instructions given to these two groups did indeed focus their attention on the spatial-temporal relationship between the limbs, then it could be argued that such an "internal" focus of attention was disruptive to acquiring a new coordination pattern. This seemed to be the case because of the bias in performance towards the anti-phase pattern during practice and in the transfer test.

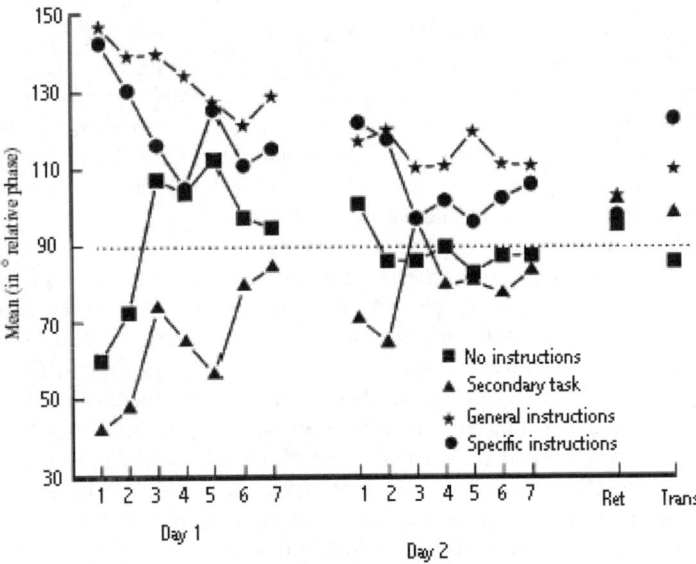

Figure 3. Effects of task instructions on mean performance in the acquisition, retention and transfer of a new bimanual coordination pattern (adapted from Hodges and Lee, 1999).

Another study (Hodges & Frank, 2000) supported this general finding using a different research strategy. In this study, the learner's internal or external focus of attention was influenced by a video of a model performing the coordination pattern (also a 90° relative timing), which the participants watched prior to physically practicing the new coordination task. Participants in two groups were told to focus either on the model's augmented feedback that was shown in the video, or the relationship between what the model's upper limbs were doing and the concurrent augmented feedback. Both of these conditions were considered to focus attention "externally" compared to another group of participants who were not given these additional instructions and received only a demonstration of the 90° pattern (the "internal" focus instructions). Once again, it was this internal focus of attention group that were to slowest to improve performance on this task over two days of practice and also the poorest in several tests of retention and transfer. A similar, detrimental influence of directing the participant's attention to the actions and feedback of the limbs has also recently been shown by Wenderoth et al. (2002). Again, there appeared to be an increased intrusion or bias toward the inherently stable patterns (in-phase and anti-phase) when participants focused their attention on their limbs compared to the augmented visual feedback presented on the computer monitor.

2.3 Discussion of section 2: On the role of intention in learning a new coordination pattern

Focusing attention on the relationship between the moving limbs (an "internal" focus) appears to have a similar effect as discussed in the previous section. The movement pattern becomes less of a focus than the individual movements that comprise the pattern. It would appear that the coordination pattern represents an abstract problem to which attention should be addressed in order to facilitate not only the performance of inherent pattern, but also for the learning of new patterns. In light of these findings it is interesting to recall another quote from Bernstein regarding the putative role of focus of attention during learning.
"Consciously watching the movements of a teacher and intent attention towards one's own movements make sense only at the beginning of the process of skill development, when the motor composition of the skill is being defined." (p. 203)
The findings presented in the previous section do not entirely support Bernstein's prediction, since the individuals in these experiments were novices to the coordination task and still did not benefit, even early in practice, from an internal focus of attention. It could be however, that bimanual coordination represents a special class of actions that have inherent stable modes that compete with the learning process in the attempt to acquire a new coordination pattern. Perhaps by focusing "internally" these inherently stable modes of coordination ironically make the acquisition of a new pattern *more* difficult to achieve. In such a case, the learner is not really a "novice" at the task, because two stable patterns of coordination are already in the repertoire. The performer knows how to perform a bimanual coordination pattern with conscious attention directed at the pattern (externally) and attempts to direct one's attention internally might disrupt a more natural behavior. Again, it is the coordination pattern, not the movements that comprise the pattern, to which attention seems to be more suitably directly.

3. Concluding comments

George Miller's admonition to graduate students about the role of intention can be food for thought for those interested in understanding coordination at a level that includes the deterministic behavior of the individual. Although intrinsic dynamics has maintained a very popular presence in the research literature for about two decades, it has done so largely without much concern for what the individual was intending to do. Different terms have been used in the literature and vary widely in their meanings: but it is clear that intention (or will, consciousness, strategy, volition, goal-directedness, etc.) appears to serve a critically important role in determining how stable coordination patterns are performed and how new patterns are learned. The study of coordination that excludes the role of intention appears to reveal only a subset of information about a very complex relationship.

References

Almeida QJ, Wishart LR, Lee TD (2002) Bimanual coordination deficits with Parkinson's disease: The influence of movement speed and external cueing. Movement Disord 17, 30-37

Amazeen EL, Amazeen PG, Treffner PJ, Turvey MT (1997) Attention and handedness in bimanual coordination dynamics. J Exp Psychol Human 23, 1552-1560

Bernstein NA (1996) On dexterity and its development. In: Latash ML, Turvey MT (eds.) Dexterity and its development. Erlbaum, Mahwah, NJ

Byblow W, Summers JJ, Lewis GN, Thomas J (2002) Bimanual coordination in Parkinson's Disease: Deficits in movement frequency, amplitude, and pattern switching. Movement Disord 17, 20-29

Byblow WD, Summers JJ, Semjen A, Wuyts IJ, Carson RG (1999) Spontaneous and intentional pattern switching in a multisegmental bimanual coordination task. Motor Control 3, 372-93

Carson RG, Byblow WD, Abernethy B, Summers JJ (1996) The contribution of inherent and incidental constraints to intentional switching between patterns of bimanual coordination. Hum Movement Sci 15, 565-589.

Fagard J (1994) Manual strategies and interlimb coordination during reaching, grasping, and manipulating throughout the first year of life. In: Swinnen SP, Heuer H, Massion J, Casaer P (eds.) Interlimb coordination: Neural, dynamical, and cognitive constraints, Academic Press, San Diego, 461-490

Fontaine RJ, Lee TD, Swinnen SP (1997) Learning a new bimanual coordination pattern: Reciprocal influences of intrinsic and to-be-learned patterns. Can J Exp Psychol 51, 1-9

Greene LS, Williams HG (1996) Aging and coordination from the dynamic pattern perspective. Ferrandez AM, Teasdale N (eds.) Changes in sensory and motor behavior in aging. Elsevier, Amsterdam, 89-131

Hodges NJ, Franks IM (2000) Attention focusing instructions and coordination bias: Implications for learning a novel bimanual task. Motor Control 19, 843-869

Hodges NJ, Lee TD (1999) The role of augmented information prior to learning a bimanual visual-motor coordination task: Do instructions of the movement pattern facilitate learning relative to discovery learning? Brit J Psychol 90, 389-403

James W (1890) Principles of psychology. Harvard University Press, Cambridge

Kelso JAS (1984) Phase transitions and critical behavior in human bimanual coordination. Am J Physiol-Reg I 15, R1000-1004

Kelso JAS (1995) Dynamic patterns: The self-organization of brain and behavior. Cambridge, MIT Press, MA

Kelso JAS, Fink PW, DeLaplain CR, Carson RG (2001) Haptic information stabilizes coordination dynamics. Proc R Soc Lond 268, 1207-1213

Kurtz S, Lee TD Perceptual-motor relationships in learning and transfer of a new coordination pattern. Manuscript in preparation

Lee TD, Blandin Y, Proteau L (1996) Effects of task instructions and oscillation frequency on bimanual coordination. Psychol Res 59, 100-106

Lee TD, Swinnen SP, Verschueren S (1995) Relative phase alterations during bimanual skill acquisition. J Motor Behav 27, 263-274

Miller GA (1986) Interview with George A Miller In: Baars BJ (ed.) The cognitive revolution in psychology. Guilford Press, New York, 199-222

Monno A, Chardenon A, Temprado JJ, Zanone PG, Laurent M (2000) Effects of attention on phase transitions between bimanual coordination patterns: A behavioral and cost analysis in humans. Neurosci Lett 283, 93-96

Peters M (1994) Does handedness play a role in the coordination of bimanual movement? In: Swinnen SP, Heuer H, Massion J, Casaer P (eds.) Interlimb coordination: Neural, dynamical, and cognitive constraints. Academic Press, San Diego, 595-615

Scholz JP, Kelso JAS (1990) Intentional switching between patterns of bimanual coordination is dependent on the intrinsic dynamics of the patterns. J Motor Behav 22, 98-124

Serrien DJ, Swinnen SP, Stelmach GE (2000) Age-related deterioration of coordinated interlimb behavior. J Gerontol B-Psychol 55, 295-303

Smethurst CJ, Carson RG (2001) The acquisition of movement skills: Practice enhances the dynamic stability of bimanual coordination. Hum Movement Sci 20, 499-529

Smethurst CJ, Carson RG (in press) The effect of volition on the stability of bimanual coordination. J Motor Behav

Swinnen SP (2002) Intermanual coordination: From behavioural principles to neural-network interactions. Nat Rev Neurosci 3, 350-361

Swinnen SP, Lee TD, Verschueren S, Serrien DJ, Bogaerds H (1997) Interlimb coordination: Learning and transfer under different feedback conditions. Hum Movement Sci 16, 749-785

Swinnen SP, Verschueren SMP, Bogaerts H, Dounskaia N, Lee TD, Stelmach GE, Serrien DJ (1998) Age-related deficits in motor learning and differences in feedback processing during the production of a bimanual coordination pattern. Cogn Neuropsychol 15, 439-466

Temprado JJ (2003) A dynamical approach to the interplay of attention and bimanual coordination. This volume

Temprado JJ, Zanone PG, Monno A, Laurent M (1999) Attentional load associated with performing and stabilizing preferred bimanual patterns. J Exp Psychol Human 25, 1579-1594

Temprado JJ, Chardenon A, Laurent M (2001) Interplay of biomechanical and neuromuscular constraints on pattern stability and attentional demands in a bimanual coordination task in human subjects. Neurosci Lett 303, 127-131

Turvey MT (1990) Coordination. Am Psychol 45, 938-953

Verschueren SMP, Swinnen SP, Dom R, De Werdt W (1997) Interlimb coordination in patients with Parkinson's Disease: Motor learning deficits and the importance of augmented feedback information. Exp Brain Res 113, 497-508

Wenderoth N, Bock O (2001) Learning of a new bimanual coordination pattern is governed by three distinct processes. Motor Control 5, 23-35

Wenderoth N, Bock O, Krohn R (2002) Learning a new bimanual coordination pattern is influenced by existing attractors. Motor Control 6, 166-182

Wishart LR, Lee TD, Cunningham SJ, Murdoch JE (2002) Age-related differences and the role of vision in learning a bimanual coordination pattern. Acta Psychol 110, 247-263

Wishart LR, Lee TD, Murdoch JE, Hodges NJ (2000) Aging and bimanual coordination: Effects of speed and instructional set on in-phase and anti-phase patterns. J Gerontol B-Psychol 55, 85-94

Wulf G, Prinz W (2001) Directing attention to movement effects enhances learning: A review. Psychon B Rev 8, 648-60

Yamanishi J, Kawato M, Suzuki R (1980) Two coupled oscillators as a model for the coordinated finger tapping by both hands. Biol Cybern 37, 219-225

Zanone PG, Kelso JAS (1992) Evolution of behavioral attractors with learning: Nonequilibrium phase transitions. J Exp Psychol Human 18, 403–421

Zanone PG, Kelso JAS (1997) Coordination dynamics of learning and transfer: Collective and component levels. J Exp Psychol Human 23, 1454–1480

Zanone PG, Monno A, Temprado JJ, Laurent M (2001) Shared dynamics of attentional cost and pattern stability. Hum Movement Sci 20, 765-789

Searching for (Dynamic) Principles of Learning

Pier-Giorgio Zanone and Viviane Kostrubiec

EA-2044 "Acquisition et Transmission des Habilités Motrices", Université Paul Sabatier – Toulouse III, 118 route de Narbonne, 31062 Toulouse Cedex 04, France

In order to provide a comprehensive and predictive framework for learning and memory, a dynamical pattern theory seeks for very general laws and principles that determine stability and change of behavioral patterns. In the nineties, learning was defined as the emergence of a new stable behavioral pattern involving the alteration of the entire layer of underlying dynamics. Twelve years after, we attempt to evaluate what new insights this approach may afford. After a brief outline of a dynamic theory of learning, we propose three generic principles underlying learning, coming from an overview of experimental work on bimanual coordination and pattern perception: a principle of symmetry conservation, a principle of distance, and a principle of time scales. Throughout this first round of research, a deep question lingers as to the possible existence of two routes to learning. Future research has to establish whether they correspond to two levels of behavioral organization, a metric and a topological level, discerned by Bernstein.

1. Introduction

It is quite easy to notice or assess an improvement in motor skills with practice. Nevertheless, it is pretty harder to explain how such an improvement is brought about with learning. The oldest idea is that learning corresponds to a gradual accumulation of progress with repeated experience, as reflected by a classical monotonically evolving learning curve (Logan, 1988). This intuitive assumption, however, is in sharp contradiction with a most obvious and overlooked experimental evidence indicating that learning exhibits various types of change over time. The bulk of support for a monotonic law is based on data averaged across trials and individuals. Yet, the shape of the individual functions does not have to adopt the same shape as the group mean curves, as shown by recent reevaluations of the issue (Palmeri, 1999; Anderson, 2001; Newell et al., 2001; Haider and Frensch, 2002; Heathcote et al., 2002). In fact, individual learning curves do show sudden changes in slope, exhibiting plateaus, periods of increase, or even periods of regression (Adams, 1987). The same types of evolution are also exhibited regarding the performance of memory as captured by recall tests (Anderson and Schooler, 1991).

Now, what does that mean and what does it imply? The issue of variability has mostly been evacuated from the scope of (motor) research, insofar as learning was viewed as discrete memory traces accumulating into and retrieved from a passive storehouse, through some kind of universal processes of storage and recovery. Intra-subject variability began to make sense when it became obvious that memory is an active process, engaged in a specific, purposeful activity (Conway, 1991; Koriat and Goldsmith, 1996; Glenberg, 1997). A real eye opener was the argument that general models of memory could not be attained, because generic situations and mechanisms of storage and recovery exist only in theory. In reality, there are no such things as a general recovery or a universal structure of memory adapted to all situations. Subjects are involved in specific perceptuo-motor tasks and must reorganize their memory in a useful, task-dependent fashion all the time. This assumption implies that what may be general are the principles governing such a reorganization. From this standpoint, plateaus, abrupt shifts, spontaneous progress, and regressions may be viewed as the hallmarks of the very reorganization process.

In the literature on learning and recall, reorganization of memories is not a new notion. Harking back to Bartlett (1932), Koffka (1935), Neisser (1967) and Piaget and Inhelder (1976), numerous empirical and theoretical studies have attempted to assess the phenomenon. Most illuminating evidence comes from studies on spontaneous variations in memory performance after the end of practice. Of particular interest is the spontaneous improvement in performance revealed by Ballard (1913) and Ward (1937) and thoroughly investigated in studies on consolidation (Brashers-Krug et al., 1996; Shadmehr and Holcomb, 1997), as well as the spontaneous recovery of forgotten skills documented by Pavlov (1932; Buschke, 1974). To account for such findings, it was argued that memory trace does not simply disintegrate over time, but rather undergoes various processes of reorganization, which may lead to distortions or even to intrusions, in which subjects wrongly recall information that had not been presented (Loftus and Palmeri, 1974; Neisser, 1981; Alba and Hasher, 1983; Estes, 1997; Schacter et al., 1998). This is a far cry from a view of learning as a mere process of accumulating memories.

In line with the ideas proposed by Bartlett (1932) and Neisser (1967), memory models have emphasized the active role of remembering in creating a meaningful and organized representation of the past. This classical approach, however, deals with an end-of-learning state of affairs and operates on symbols, at a level of abstraction that hides the details as to how these reorganizations are generated and how the process evolves in real time. It is one think to attribute end-of-learning states to reorganization of memory symbols evolving in an idealized space disconnected from the environment, a space in which time is merely reduced to the ordered sequence of occurring events. It is an altogether different matter to explain how such a reorganization may be a logical consequence of rate-dependent laws that govern a real organism, constrained by its proper perceptuo-motor abilities and by specific task demands. Our contention is that a new organizational state cannot to emerge unless the old one has lost *stability*. Therefore, stability is a crucial concept that yields an entry point into how reorganizations of memory with learning may be studied.

An operational approach to identifying principles leading to gain and loss of stability is provided by a dynamical pattern theory (Kelso, 1995). This framework stems from physical theories of open, non-equilibrium systems (Haken, 1983) and shares some insights and notions with ecological psychology (Turvey and Kugler, 1984). It also provides mathematical tools that help capture how action-perception patterns emerge, stabilize, and change in real time. In this view, organisms are open to environmental influences, so that action is *lawfully* related to the incoming perceptual information (Turvey et al., 1981; Turvey and Shaw, 1995; Schöner et al., 1998). Learning implies that preexisting perceptuo-motor linkages be relinquished at times, so that the system is free for establishing novel and different relationships between action and perception. A major endeavor is then to unravel what general principles may underlie these reorganizations.

This chapter presents recent findings about learning-induced modifications of perceptuo-motor couplings, so-called "coordination dynamics". After a brief outline of the dynamics of learning (Section 2), empirical evidences will be provided suggesting generic underlying principles. Three principles will be put forth: the principle of symmetry conservation (Section 3), the principle of distance (Section 4), and the principle of time scale (Section 5). Throughout the chapter, we shall present a rationale for an experimental strategy and discuss a key issue pertaining to the routes that learning may take. Finally, we shall discuss new findings that might open new paths of investigation (Section 6).

2. Learning dynamics in bimanual coordination

A perceptuo-motor *pattern* is defined operationally as a stable and reproducible relationship between the components of a biological system under specific task requirements. Various influences can alter preexisting perceptuo-motor patterns as learning proceeds, such as environmental, memory, or intentional requirements. These demands are envisaged as specific *constraints* acting on the perceptuo-motor linkages and attracting the behavioral pattern toward a required value (Schöner and Kelso, 1988c). A critical hypothesis is that when memory for a specific pattern strengthens, the entire perceptuo-motor relationship changes, and not simply the particular pattern being practiced (Schöner et al., 1992). This assumption entails a specific experimental rationale, in order to study and formalize such a learning-induced alteration. The first experimental prerequisite is to probe the ensemble of existing perceptuo-motor relationships before learning, revealing initially stable patterns. Next, the pattern to be learned must be selected in reference to the capabilities of each individual learner before practice, so that it is really new, that is, not preexisting. Finally, at the end of the learning procedure, the whole perceptuo-motor relationship must be probed again, in order to assess the modifications due to learning (Zanone and Kelso, 1992). Such alterations of perceptuo-motor relationships may be conceived of as memory reorganization.

2.1 Dynamical approach to learning

A strength of a dynamical approach bears on conceptual and experimental strategies instrumental to discovering the *laws* that govern changes in perceptuo-motor relationships. These laws may be captured by assessing the evolution of the *stability* of the perceptuo-motor pattern, defined as the capacity of the system to damp down perturbations coming from external or internal sources. A fairly stable pattern can thus persist for an extended period of time in the face of continuously fluctuating influences. In contrast, when stability is lost, a small perturbation may lead to the emergence of a new stable behavioral state (Haken et al., 1995). In this light, memory is equated to a stable perceptuo-motor pattern, and learning to the stabilization of a new, so far unstable pattern. Forgetting means loss of stability of learned/stabilized patterns. The crux of a dynamic approach to learning lies in evaluating the evolution of stability empirically and in capturing it formally. This should lead to non-trivial predictions, directly testable through experimentation (Schöner, 1989).

In order to assess stability, several measures are available, in particular standard deviation (SD) or relaxation time (τ_{rel}). Relaxation time indicates the time needed for an evolving system, characterized by a state vector $\mathbf{q}(t)$ defined in a n-dimensional space R^n, to return to a stable state $\mathbf{q}_0$ after a small perturbation has driven it away. The shorter τ_{rel} is, the more stable the state. Relaxation time turns out to be independent of the magnitude of the perturbation and is related to SD (see below). SD reflects stochastic perturbations acting on $\mathbf{q}(t)$ and causing the system to fluctuate around $\mathbf{q}_0$. Both indexes of stability increase as stability decreases (Schöner and Kelso, 1988a).

The first step toward a dynamical model of perceptuo-motor learning is to focus on relevant behavioral variables by means of scrutinizing the evolution of the pattern stability. Loss of stability can lead to abrupt, qualitative changes in $\mathbf{q}(t)$, called *bifurcations* or *phase transitions*. When $\mathbf{q}(t)$ undergoes a bifurcation, particular variables, $\phi_i(t)$, may be revealed as they loose stability. Such variables organize all microscopic components of the system, $\mathbf{q}(t)=(q_1,...,q_n)$, giving rise a coherent pattern at the collective level. Thus, $\phi_i(t)$ are said to govern the macroscopic behavior of $\mathbf{q}(t)$ and constitute appropriate variables to capture the dynamics at a collective level. Because $\phi_i(t)$ represent k orthogonal axes of the system, this strategy leads to a highly abstract and low-dimensional model defined in $R^k \subset R^n$, independent from the microscopic details of the component behavior (Haken, 1993; Haken et al., 1993; Haken et al., 1995). Therefore, a dynamical approach (to learning) allows describing the system evolution (with learning) without addressing explicit propositions with regard to the neural or biological implementation of the process.

Formally, a dynamical model refers to an *equation of motion* expressing the rate of change of $\phi(t)$. Such *dynamics* often takes the form of the first derivative of ϕ with regard to t, denoted as $d\phi/dt$. Behavior over time is determined by the evolution of *asymptotically stable states* or *stable fixed points* in the *local* stability regime. A local solution, ϕ_0, is said to be asymptotically stable if, after a small perturbation, the system returns to its former value as time proceeds ($t \rightarrow +8$). In other words, a stable state is a solution of the differential equation, such that $d\phi/dt$

$$\tau_{rel} = \frac{1}{|\lambda|} \quad (2.1.1)$$

= 0 and such that the derivative of $d\phi/dt$ with respect to ϕ is negative at this point. Therefore, this second derivative constitute an index of stability, λ, which is related to relaxation time as in equation (2.1.1).
This index, of course, proves to be related to standard deviation about ϕ_0 (Schöner et al., 1986) equation (2.1.2).

$$SD_{\phi_0} = \sqrt{\phi \frac{Q}{2|\lambda|}} \quad (2.1.2)$$

A stable fixed point is also coined as an *attractor*, because it biases all nearby trajectories toward the stable value. The region from which behavioral trajectories return toward an attractor is called a *basin of attraction*. Visually, an attractor is easily determined when the rate of change, $d\phi/dt$, is plotted as a function of ϕ. In this space, a point at which $d\phi/dt$ equals to zero, ϕ_0, specifies a so-called *fixed point* or *stationary solution*. When the rate of change has a negative slope at ϕ_0, the stationary solution depicts an attractor of the system (Figure. 1). The steeper the slope is, the faster the system returns to this point and the stronger the attractor.

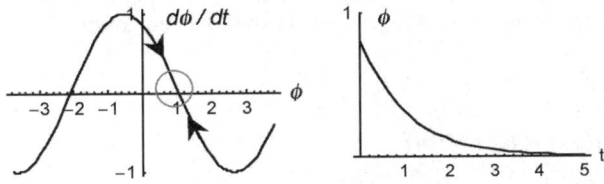

Figure 1. Left panel: rate of change $d\phi/dt$ is plotted as a function of the behavioral variable $\phi_i(t)$ The zero-crossing of this function indicates a fixed point corresponding to a stable solution, because the slope of $d\phi/dt$ is negative. Right panel: convergence of a behavioral trajectory toward the attractor.

To spell out a dynamical model of learning, an adequate strategy is to map the behaviorally stable patterns onto attractors of the equation of motion, $d\phi/dt$. Then, a task requirement also has to be mapped onto an attractor for the behavioral variables at a given value. The process of learning is thus visualized by the deformation of $d\phi/dt$ taking the form of the emergence of an attractor at the task requirement value (Zanone and Kelso, 1992). The best rendition of the momentary state of learning is to plot the error between the solutions of $d\phi/dt$ and a given task requirement across the whole set of task requirements (Figure 2, left panel). When an intersection of the error curve with the abscissa shows a negative slope, it indicates an attractor. Displaying SD as a function of the required phase provides additional information, as an attractor corresponds to a point at which SD is lowest (Figure 2, right panel). This complete picture is referred to as a *probe* or a *scan* (see Zanone and Kelso, 1992 for details).

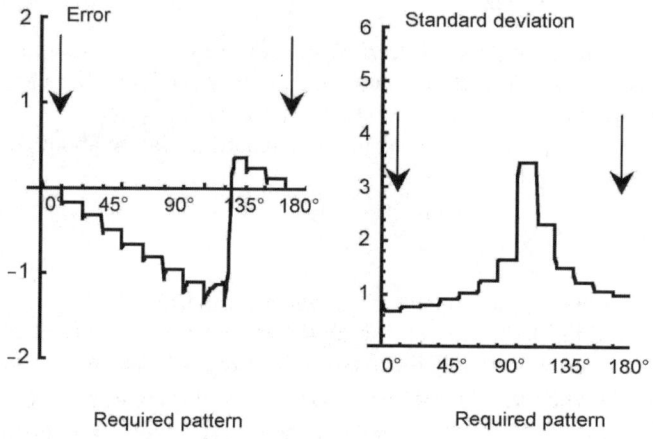

Figure 2. Error between the performed and the required behavioral pattern and the associated SD (left and right panel, respectively), as a function of the task requirement. Thirteen equally spaced task requirements, denoted in degrees, are presented in ascending order every 5 time units. Zero-crossing of the error with negative slope and minimal SD reveals attraction. Arrows indicate attractors at two task requirements 0° and 180°.

2.2 Bimanual coordination

Because the best experimental situations are the simplest, the original idea was to study a task-specific system that exhibits only few stable patterns prior to the learning. A good case is bimanual coordination, that is, coordination between homologous limbs moving periodically: Behaviorally, only two patterns of coordination are spontaneously stable, *in-phase* and *anti-phase* (Kelso, 1984). The in-phase pattern involves symmetrical motion of the limbs in opposite directions, as a result of the activation of homologous muscles, whereas the anti-phase pattern refers to motion in the same direction, with simultaneous activity of antagonist muscles (Figure 3). The in-phase pattern appears to be more stable than the anti-phase pattern, and a transition from the latter to the former occurs when the oscillation frequency increases beyond a critical threshold. Previous work established that a relevant collective variable to capture such dynamics is *relative phase*, ϕ, between the limb motion, yielding $\phi = 0°$ for in-phase and $\phi = 180°$ for anti-phase (see Figure 3). Zanone and Kelso (1992) showed that the eventual stabilization of a new learnt pattern with practice may take the form of abrupt shifts in performance. This process is accompanied by an increase in accuracy and stability (Hodges and Franks, 2000; Hodges and Franks, 2001), as well as by a decrease in relaxation time (Wenderoth and Bock, 2001). In addition, the process affects spontaneously stable patterns, leading to their temporary destabilization (Lee et al., 1995; Fontaine et al., 1997; Wenderoth et al., 2002).

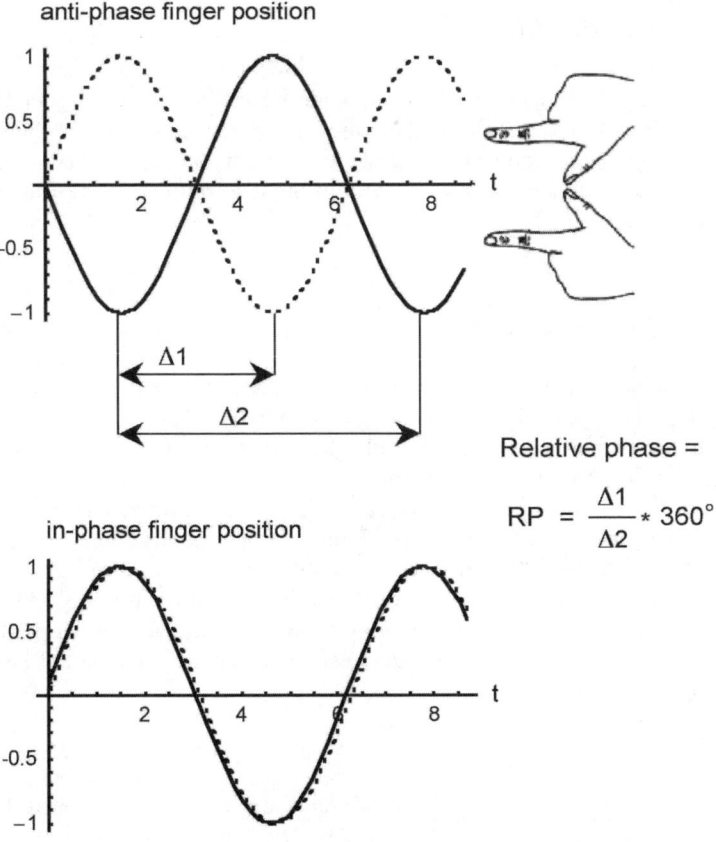

Figure 3. In-phase and anti-phase behavioral patterns. The plots display right (dotted line) and left (solid line) finger position as a function of time. The top panel presents an anti-phase pattern, as well as the method of relative phase calculation The bottom panel displays an in-phase pattern.

This behavioral picture was formally captured in a model by Schöner and Kelso (1988b; Schöner, 1989), extending the original HKB model of bimanual coordination (Haken et al., 1985). Given the two stable attractors at 0° and 180°, the model HKB postulates a bimanual coordination dynamics developed as a truncated Fourier series of the second and third sine terms. Because this function depicts the dynamics of spontaneously stable states, we shall refer below to it as an *intrinsic dynamics*, $f_{intr}(\phi,t)$ equation (2.2.3),

$$\frac{d}{dt}\phi(t) = f_{intr}(\phi,t) = -a\sin[\phi(t)] - 2b\sin[2\phi(t)] + \sqrt{Q}\xi(t) \qquad (2.2.3)$$

where the last expression denotes stochastic fluctuations due to a Gaussian white noise ξ of strength Q. A more intuitively understandable picture may be given by the potential function $V(\phi,t)$ of the dynamics, in which the system is "viewed" as a particle evolving in a landscape following gravitational forces. Technically, the potential is defined as the negative integral of the deterministic part of the intrinsic dynamics equation (2.2.4),

$$V(\phi,t) = -\int f_{intr}(\phi,t)\,d\phi = -a\cos[\phi(t)] - 2b\cos[\phi(t)] \qquad (2.2.4)$$

For positive a and b pertaining, for instance, to the oscillation frequency, this 2π-periodic, $V(\phi,t) = V(\phi + 2\pi,t)$, and symmetric, $V(\phi,t) = V(-\phi,t)$, potential function $V(\phi,t)$ defines a dynamical landscape with two attractors rendered by its minima, or wells, giving rise to *bistable* dynamics. When random perturbations push the system away from such an attractive state along the potential curve, the system relaxes then to an attractor at a speed proportional to its slope. In a flattish potential, the system would relax very slowly to its equilibrium value, whereas when the potential well is steep and narrow, relaxation time is short. With changing a and b parameters, the well may vanish, so that every small perturbation can move the system toward a more stable state, leading to a change of behavioral pattern (Figure 4).

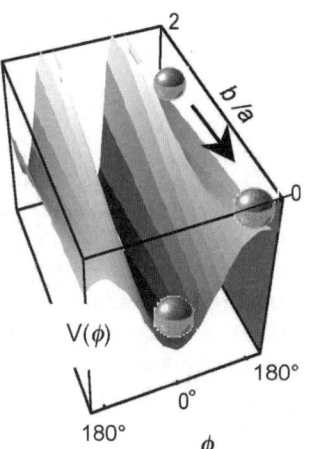

Figure 4. Potential function $V(\phi,t)$. When the 180° potential well is deep and narrow (top particle), random perturbations are quickly damped down and relaxation is fast. When the attractor vanishes, relaxation time becomes longer, until stability is lost (middle particle): a small perturbation makes the system relax toward a more table state, 0° (bottom particle). The change of the potential is obtained by varying b/a from 2 to 0, with a remaining at 1.

In this framework, learning is expressed as the emergence of a novel stable state close to the task-required value, Ψ_{env}, which attracts the behavioral variable with a strength defined by c_{env}. To account for learning, a new degree of freedom must be introduced, a memory variable, Ψ_{mem}. In the simplest if unrealistic case, only one pattern can be learned. The task requirement Ψ_{env} and the related memory Ψ_{mem} are assumed to act on the coordination dynamics according to equation (2.2.5):

$$\frac{d}{dt}\phi_{total}(t) = -a\sin[\phi(t)] - b\sin[2\phi(t)] - c_{env}\sin[\phi(t) - \Psi_{env}]$$
$$- \hat{c}_{mem}[\Psi_{mem}(t)]\sin[\phi(t) - \Psi_{mem}(t)] \qquad (2.2.5)$$

The strength of the memory variables is bounded by c_{mem}. In this formulation, learning a new pattern is captured through two aspects. First, the memory variable value Ψ_{mem} evolves toward the value required by the task, Ψ_{env} as in equation (2.2.6):

$$\frac{d}{dt}\Psi_{mem}(t) = -\tau_{learn}^{-1}\sin[\Psi_{mem}(t) - \Psi_{env}] \qquad (2.2.6)$$

where τ_{learn} represents the relaxation time of the attractor at the to-be-learned value. Second, the strength c_{mem} of the memory variable increases as Ψ_{mem} converges toward Ψ_{env}, according to equation (2.2.7):

$$\hat{c}_{mem}[\Psi_{mem}(t)] = c_{mem}\cos^2\left[\frac{\Psi_{env} - \Psi_{mem}(t)}{2}\right] \qquad (2.2.7)$$

In a a more realistic case, in which several memory variables Ψ_i are possible, an additional degree of freedom is introduced in order to select the memory variable that learns the task requirement. This activation variable, w_i, associated with each Ψ_i, defines the relative activation of the memory variables under the influence of a given task demand equation (2.2.8):

$$w_i(t) = \frac{|n_i(t)|}{\sum_{i=1}^{N}|n_j(t)|}, \quad 0 < w_i < 1, \quad \sum_{i=1}^{N}w = 1, \quad (i = 1,2,...,N) \qquad (2.2.8)$$

When Ψ_{env} is presented to the system, memory variables enter in competition with regard to the similarity, $m(\phi, \Psi_{mem})$, between Ψ_i and Ψ_{env}, as well as Ψ_i and ϕ equation (2.2.9):

$$m[\phi(t), \Psi_i(t)] = \tau_{rec}^{-1}\left[\frac{\cos[\phi(t) - \Psi_i(t)] + \cos[\Psi_{env}(t) - \Psi_i(t)]}{2}\right] \qquad (2.2.9)$$

The most activated w_i^* winning this competition inhibits the activation of the remaining w_i, so that they all relax to zero. This competition is shaped by the last term parametrized by r in equation (2.2.10):

$$\frac{d}{dt}n_i(t) = m[\phi(t), \Psi_i(t)]n_i - r[\sum_{j=1}^{N} n_j^2(t)]n_i(t) + Q\xi_i(t) \quad (2.2.10)$$

Such a term subtracts from each w_i the activation of all remaining w_i, thereby allowing the expression of a *cooperative action* of memories. The crux of this model is that the most activated memory variable, Ψ_i^*, is attracted toward the task requirement, while less activated variables converge in the direction of Ψ_i^* following equation (2.2.11),

$$\frac{d}{dt}\Psi_i(t) = -\tau_{rec}^{-1} w_i \sin[\Psi_i^*(t) - \Psi_i(t)] \quad (2.2.11)$$

where τ_{rec} refers to the time scale of recall, such as $\tau_{rec} \ll \tau_{learn}$. Thus, the inhibition term governs learning ($\Psi_i^* \rightarrow \Psi_{env}$), as well as forgetting by interference ($\Psi_i \rightarrow \Psi_i^*$). This model provides an operational entry point into the key issue of learning and forgetting, through the window of bimanual coordination. The entire evolution is illustrated in Figure 5 (for more examples with explanations, see Schöner 1989).
In the rest of the chapter, we shall present empirical findings that support such a view establishing a tight link between learning and memory. In particular, some experimental work of ours, strongly inspired by the above model, investigated how symmetry properties, the distance between the task requirements and intrinsic attractive states, and time scales may influence the process of learning.

3. Principle of symmetry conservation

In bimanual coordination, a critical observation is that the spontaneously stable states are invariant over the transformation $+\phi \rightarrow -\phi$. Formally speaking, symmetry is a transformation that maps an object onto itself. Because symmetry is a fruitful concept, both explanatory and predictive for pattern formation in biology and for human perceptuo-motor skills (Thompson 1961; Kelso and Jeka 1992; Turvey and Shaw 1995; Fuchs and Jirsa 2000), it is likely to play a key role in learning too. A good entry point is a statement by Mach (1902), *"in every symmetrical system every deformation that tends to break the symmetry is complemented by an equal and opposite deformation that tends to restore it"* (cited in Shaw and McIntyre, 1974). For bimanual coordination, symmetry is broken as soon as the to-be-learned pattern, specified by the environmental task, becomes a new attractive state. Thus, if some principle of symmetry conservation comes into play in learning, then practice of a new, spontaneously unstable pattern should automatically give rise to the stabilization of the symmetric, initially unstable pattern.

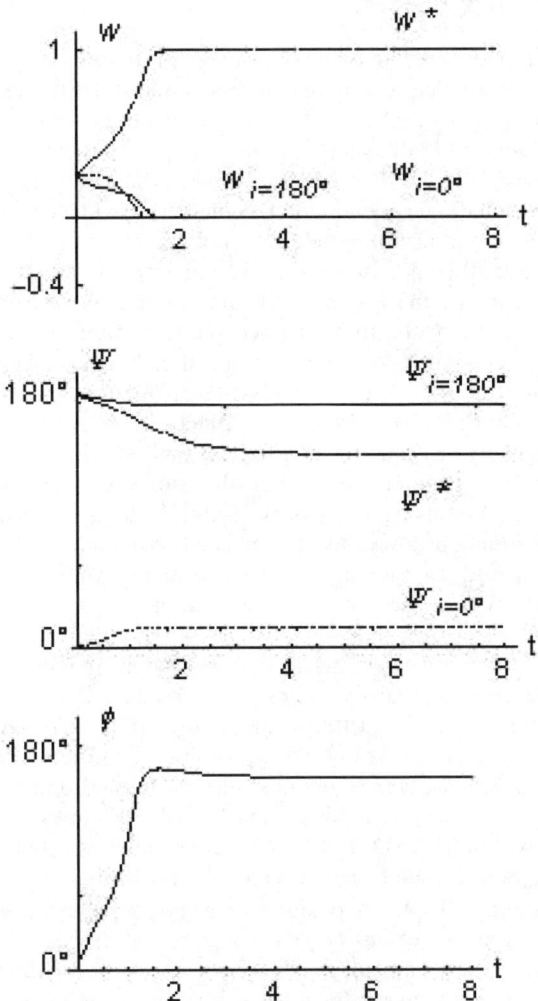

Figure 5. Simulation of learning a 135° task by an initially bistable system at 0° and 180°. Evolution of the activation variables (top panel), memory variables (middle panel), and relative phase (bottom panel), as a function of time. The initial conditions were set at $\Psi_i=0°$ (t=0)= 0°, $\Psi_i=180°$ (t=0) = 180°, and $\Psi_i^*=180°$ (t=0) = 180° for memory variables, $w_i=0°$ (t=0) = 0.30, $w_i=180°$ (t=0) = 0.30, and $w_i^* =180°$(t=0)=0.40 for corresponding activation variables, and ϕ (t=0)=0° for relative phase. The simulation shows that the memory variable initially most activated shifts toward the required value, while the remaining memory

variables drift slightly toward the most activated memory variable. The curves show general tendencies superimposed on generated data.

3.1 Transfer within an effector system

This hypothesis was tested by Zanone and Kelso (1997). In a learning procedure, subjects were asked to practice a required, to-be-learned relative phase pattern with knowledge of results given after each trial. The required relative phase was displayed by two light-emitting diodes (LED) allowing various relative phases to be generated by manipulating the time interval between the LED onsets. The task was to learn to flex each finger in temporal coincidence with the onset of the ipsilateral LED. After familiarization with the experimental set-up, the underlying dynamics was fully probed from 0° to 180° and from 180° to 360° in 24 discrete steps of 15°, thereby scanning the entire relative phase span between the in-phase and anti-phase patterns. Then, the to-be-learned phase relationship was chosen, under the caveat that it did not coincide with an already-existing coordination pattern. Thus, subjects with bistable dynamics, that is, exhibiting a stable behavior at 0° and 180°, were assigned to two groups who practiced either 90° (i.e., the left finger lagging the right by one quarter of a cycle) or 270° (i.e., the left finger leading the right by the same amount). For the subjects with "multistable" dynamics, characterized by four spontaneously stable patterns, such as 0°, 180° and 90° or 270°, the learning task was set at an unstable pattern (e.g., 135°). At the end of the training period comprising 50 practice trials, a full probe of the coordination dynamics was carried out to assess the alteration of coordination dynamics due to practice.

Analysis of the final scan revealed that the to-be-learned pattern became an attractive state of the coordination dynamics with practice. Comparison of the probes before and after practice illustrates this change (leftmost and rightmost graphs in Figure 6). The upper curves (solid lines) plot the mean error in relative phase, that is, the average difference between the performed and the required relative phase within a trial. The lower curves (dotted lines) display the corresponding mean within-trial SD. Both scores are plotted as a function of the phasing requirements, and for both curves vertical bars denote between-subject variability, encompassing ± 1 SD. A positive or negative value means that the required relative phase was overestimated or underestimated, respectively.

For the scan before learning (leftmost graphs), error is lowest for the 0° or 180° requirements, with a marked increase for intermediate values. The negative slope reflects an attraction of nearby phasing requirements to 0° and 180°. This reveals the stability of the in-phase and anti-phase patterns before learning and suggests that these patterns are attractors of underlying bistable dynamics. For the scan after learning (rightmost graphs), the error curve showed in addition a negative slope and a low SD around the 90°practiced pattern (denoted P). This overestimation of the required pattern below 90° and underestimation above it means that the pattern actually performed was biased toward 90°, indicating that the learned state has become attractive to its neighbors. Still, a remarkable finding was that the symmetry partner of the to-be-learned pattern became an attractive state too, although such a pattern had *never been practiced at all*. In Figure 6, when a 90° phasing was learned, subjects were also able to perform a 270° non-practiced phasing pattern in a stable fashion (denoted NP). These results were corroborated by Smethurst and Carson (2001), specifying that after practice, in a paradigm *à la* Kelso where the pacing frequency was progressively increased, transitions from 90° were generally to the anti-phase 180° pattern, whereas transitions from 270° were to the 90° pattern.

Such a spontaneous transfer of learning between relative phase patterns of opposite signs implies that the symmetry of the underlying coordination dynamics is preserved as learning proceeds. In other words, learning a new phase relationship occurs irrespective of any time ordering between the moving fingers, a sign that such an acquisition is fairly effector-independent. This finding provides a strong support to the basic assumption that learning operates at the abstract level of coordination dynamics, and not only at the behavioral level of performance.

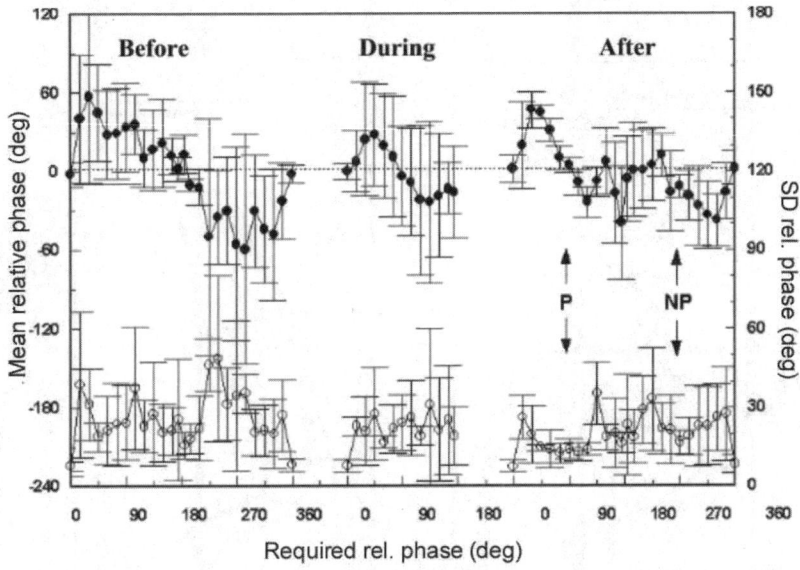

Figure 6. Within effector system transfer of learning with practice of a 90° pattern. Zero crossing with negative slope and minimal SD indicate the presence of an attractor (denoted by an arrow) at this very task requirement.

3.2 Transfer between effector systems

The question now is whether symmetry conservation also comes into play across two different end-effector systems. Kelso and Zanone (2002) investigated whether learning a novel phase relationship transfers spontaneously from arms to legs, and conversely. Six subjects were asked to learn a specific relative phase through

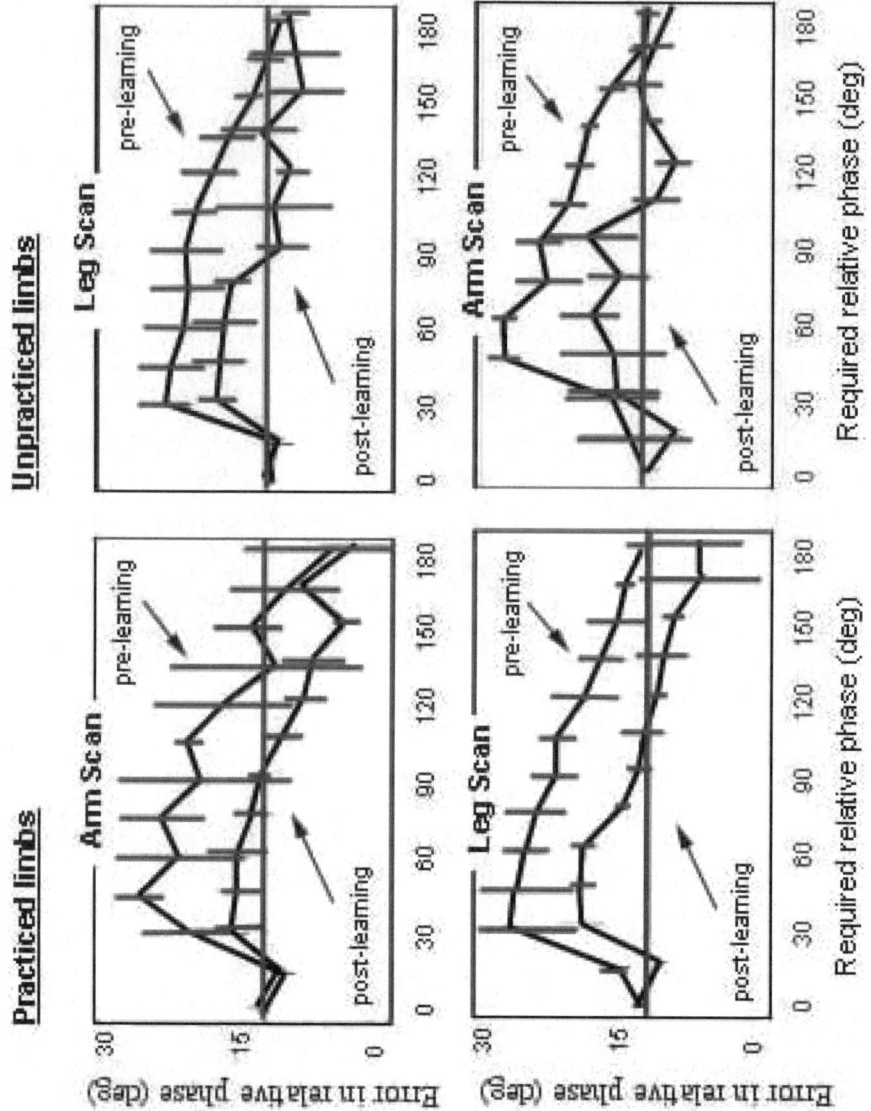

practice with either the arms or the legs. The instructions were synchronize peak

Figure 7. Across effector systems transfer of learning. Zero crossing with negative slope indicate the presence of an attractor.

motion of the limb (i.e., the top reversal point) precisely at the moment when the corresponding LED was turned on. Six control subjects were not exposed to such a practice. To assess modifications induced by learning and transfer, the coordination dynamics pertaining to the arm and leg systems were assessed before and after practice through probes for both the practiced and unpracticed limbs.

After the first scan, the to-be-learned pattern was set on an individual basis such that it did not correspond to a preexisting stable pattern. The selected new pattern was practiced with one effector system for 40 trials, with KR given at the end of each trial.

Figure 7 shows the results of the leg and arm probes carried out before and after practice, for the probes of the practiced (left column) and unpracticed limbs (right column). In the top left panel, comparison of the pre-learning and post-learning error curves (dotted vs. solid lines, respectively) reveals the changes affecting the probes for the subjects practicing a 90° relative phase with the arms. Before practice, error are lower for the 0° and 180° patterns, indicating that in-phase and anti-phase were attractive states of the underlying coordination dynamics. After practice, error curves shoved in addition a negative slope close to the learning task requirement. Such changes reveal learning, that is, the practiced pattern becomes a new stable state of the coordination dynamics. Strikingly, the same basic alterations appeared in the leg scan, although the legs *did not practice the task at all* (top right panel). These are unambiguous signs of spontaneous transfer of learning across two effector systems. A similar transfer of learning was observed for subjects practicing a 90° relative phase with the legs, as revealed in the two bottom panels. An interesting question is whether such transfer pertains to the similarity of the coordination dynamics before practice. In particular, both leg and arm dynamics were initially symmetric and were bistable at in-phase and anti-phase.

3.3 Order of pattern stabilization

The foregoing findings about spontaneous transfer imply that the HKB model must be expanded in order to account for the symmetric changes due to learning. Recall that the original model was developed as a Fourier series with the second and the third sine term ($a_n \sin(n\phi)$, with $n_2=1$ and $n_3=2$), generating stable states at 0° and 180° (cf. Eq. 2.2.4). Setting the next even terms to a nonzero value leads to the emergence of new attractors close to 90° for $n_4 \neq 0$, to 60 and 120° for $n_6 \neq 0$, and to 45° and 135° for $n_8 \neq 0$. A challenging question is whether learning gives rise to these attractors in the same order as that of the Fourier expansion, that is, a succession of higher and higher even sine terms. In other terms, an issue is whether the symmetry structure of the bimanual coordination dynamics constrains the learning process in a substantial fashion.

An operational strategy to test the influence of coordination dynamics on learning was to exert an unspecific learning force on the entire attractor layout, requiring, so to speak, the acquisition of whatever relative phase but the spontaneous in- and anti-phase patterns. In a study undertaken in our laboratory (Atchy-Dalama, 2000), subjects had to learn to move a pair of customized joystick with pronation-supination forearm movements in temporal coincidence with two LEDs displaying various phasing requirements. The study involved seven relative phases (0°, 30°,

60°, 90°, 120°; 150°, and 180°) presented five times in a random order. At the end of each training period involving all seven requirements, KR was returned to the subjects about the produced performance for each phasing. Practice sessions were held on three separate days, with a scanning probe carried out before and after each session.

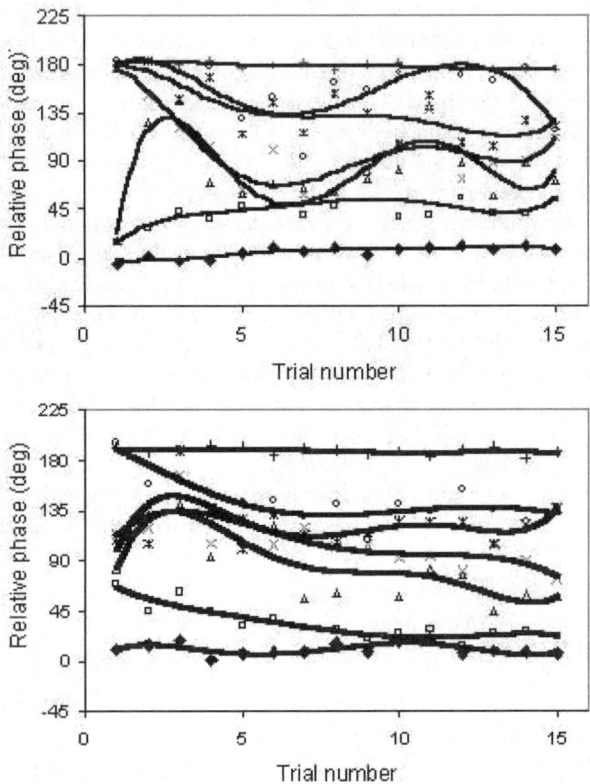

Figure 8. Ordered apparition of stable patterns with unspecific learning. Performance of the seven relative phases to be learned is plotted as a function of the probe number, along with a cubic interpolation. Progressive clustering of relative phase values suggests the formation of a basin of attraction.

A complete representation of the evolution of an individual coordination dynamics with learning is provided in Figure 8. For the each probe, the mean relative phase corresponding to the seven probed phasing requirement, from 0° to 180°, is plotted as a function of the probe number. In order to appreciate the evolution of the attractor layout, a cubic polynomial interpolation is drawn across probes for each task requirement. If no change in the dynamics occurred with practice, seven equidistant horizontal lines would be drawn. In contrast, clusters of relative phase values would suggest the apparition of a basin of attraction.

Figure 8 (top panel) shows that bistable coordination dynamics gave rise to the emergence of a new basin of attraction about 90°. This result may be formally captured by adding the fourth sine terms to the original HKB model. Multistable coordination dynamics (Figure 8, bottom panel) led to the emergence of two symmetric basins of attraction at 45° and 135°. Formally, the sixth and eighth sine terms should be inserted into original HKB model, the sixth term leading to the loss of stability of 90°.

An essential finding of this study was that, at any rate, learning exhibits some signs of symmetry preservation under such a massive, if unspecific environmental demand. This first evidence of an ordered apparition of attractors requires, however, a closer theoretical and empirical examination. Indeed, the loss of stability of 90° with learning may be explained otherwise. In multistable dynamics, this preexisting stable state might merely "shift" in the direction of the new phasing requirement, under the effect of memory variables. Thus, there are in principle two possible *routes to learning*, that is, two different ways by which the initial attractor layout may be modified so that the required pattern eventually becomes a stable state of the coordination dynamics (Zanone and Kelso, 1994). The findings presented in the next section will get us a bit further into this issue.

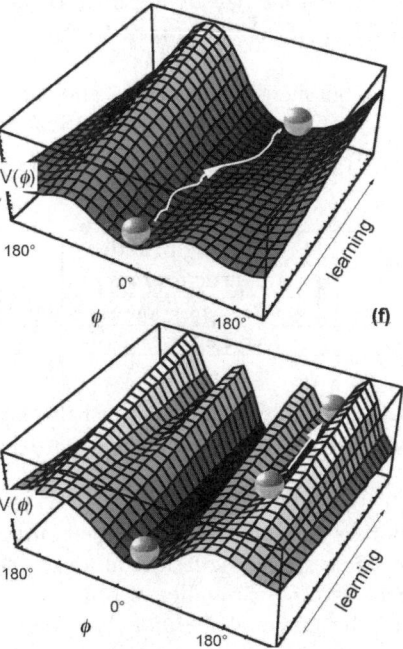

Figure 9. "Shift" and "bifurcation" routes to learning. Top panel: With the "shift"-route, there is a rapid improvement in accuracy, followed by a later increase in stability. Bottom panel: In contrast, the "bifurcation"-route leads to more homogenous increase in accuracy and stability.

4. Principle of distance

In the above section, we suggested that there are two manners to perform an initially unavailable coordination pattern close to the task requirement (Zanone and Kelso, 1994). The first mechanism is to "create" a new basin of attraction by directly altering the intrinsic dynamics (Figure 9, top panel). The findings by (Zanone and Kelso, 1992) on learning illustrate this type of change, characterized by a bifurcation. The second mechanism is to "shift" a preexisting attractor and to "position" it at the required value (Figure 9, bottom panel). Our goal now is to provide empirical support for the two types of change in the underlying coordination dynamics with practice.

To capture these two routes to learning experimentally, a key question arises as to how a dynamical system may "choose" between the two routes. A first possibility is that such a choice may depends on the distance, Δ, between the preexisting memory variables and the task requirement. A "shift"-route should be observed when the task requirement is fairly close to a nearby stable state. Conversely, the emergence of a new attractor following a "bifurcation"-route should occur when the task requirement is rather distant from preexisting attractors. In sum, the principle of distance states that *the type of alteration of the coordination dynamics with learning depends on the distance between the preexisting attractors and the pattern to be learned.*

Distance, Δ, is an accessible metric empirically, and such a principle of distance led to a rich set of hypotheses. According to (Schöner 1989)'s model, two contradictory conclusions may be drawn for during and after practice, as summarized in Tab. 1.

Distance	During practice *practice task*	After practice *prompting task*
Small	low error, low SD	large error, high SD
Big	large error, high SD	low error, low SD

Table 1. Predicted evolution of error and SD during and after practice, as a function of the distance between spontaneously stable states and task requirements, after (Schöner 1989)'s model.

On the one hand, if during the practice the required attractor Ψ_{env} coincides with one of the preexisting stable states, the task requirement stabilizes the preexisting pattern, so that the error and the variability of produced pattern will be minimal. This interplay between the task and the underlying coordination dynamics was coined as a *cooperation regime* (Zanone and Kelso, 1992; Schöner and Keslo, 1988). In contrast, when the preexisting and required states are distant, the produced pattern deviates from the requirement toward the closest stable pattern, so that both the error and variability are large. This constitutes a *competition regime*. Hence, when the distance, Δ, is small, accuracy and stability of to-be-learned pattern should be high, and conversely for large Δ (cf. Eq. 2.2.5). On the other hand, after practice, distance has a main effect on forgetting by interference, so that the larger the distance is, the weaker the interference (cf. Eq. 2.2.11). Thus, accuracy with regard to the newly learned pattern and stability should still be large for memory variables far from the task requirement, and conversely for small Δ.

4.1 Learning and interference

The foregoing hypotheses were systematically explored by Kostrubiec and Zanone (submitted). Subjects had to learn three relative phase patterns consecutively, namely $\phi = 90°$, $\phi = 135°$, $\phi = 158°$, corresponding to $\Delta = 90°$, $\Delta = 45°$, and $\Delta = 22°$ apart from the two neighboring and equidistant stable patterns, respectively. For each relative phase, practice trials were interspersed with so-called "prompting" tasks. In a practice trial, participants had to produce the to-be-learned pattern, as displayed by the LEDs, for 20 s, immediately followed by a complete KR. In the prompting task, instructions were to first observe the LEDs displaying the two preexisting stable patterns closest to the to-be-learned pattern, and then perform it from memory.

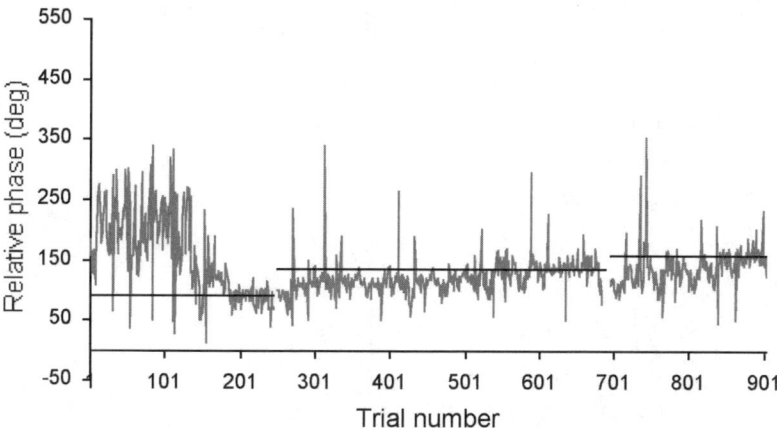

Figure 10. Evolution of the performed relative phase during learning a 90°, 135°, and 158° relative phase consecutively. Horizontal lines indicate the successive task requirements.

This procedure was aimed to assess the current state of nearby relative phase patterns already memorized, while the stabilization of a new pattern with practice was underway. More specifically, the 0° and 180° patterns were scrutinized while 90° was practiced, the 90° and 180° patterns were so for the 135° practice trials, and the 135° and 180° patterns for the 158° trials. After eight seconds, the LEDs were turned off; and a beep prompted that the withdrawn target value be performed from memory for 20 s, in the absence of any information. The criterion for a completed learning was that both low error (=12°) and low variability (=20°)

were reached for three trials on a row, thereby equating the performance level between subjects and learning tasks.

In line with our predictions, our findings revealed that the 90° pattern, distant from preexisting attractors, was significantly less accurate than the 135° and 158° patterns. A novel finding was that two routes to learning were observed as learning proceeded. Figure 10 displays the performed relative phase for each relative phase requirement, as a function of the practice trials. For the 90° task requirement, the path toward 90° involved a bifurcation, whereas a "shift"-route occurred for 135° and for 158°.

An interesting consequence is that the route that learning process takes has an impact on its rate. Assuming that the number of trials needed to reach the learning criterion is a fair estimator, the learning rate turned out to be significantly larger for 135° than for 90° and 158°. The "shift"-route happened to be slower than that the "bifurcation"-route.

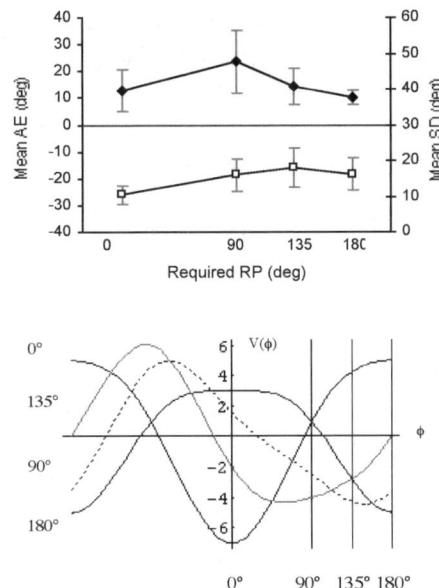

Figure 11. Top panel: mean AE (upper curve) and SD (lower curve) as a function of the required RP in the prompting task. Bottom panel: simulation of the results by a coordination/competition model.

Let us consider now the interference issue. In order to test the effects of the distance Δ on the distortion of extant coordination patterns (i.e., either intrinsic or newly learned), we compared accuracy and stability in the prompting task, as illustrated in the upper and lower curves of in the top panel of Figure 11, respectively. Accuracy was significantly lower for the learned patterns (90° and

135°) than for the initially stable patterns (0° and 180°), and there was a reliable decrease in accuracy between the learned 135° and 90° pattern. Of particular interest was that the distance Δ affected the distortion of already learned patterns, namely, interference, in tight accordance with the predictions of the competition-cooperation model presented in the bottom panel of Figure 13. The simulation shows that the 0° attractor (the left black curve) is more stable and more accurate than the other attractors, whereas 0° and 180° (right black curve) are more accurate than 135°, while 135° is more accurate than 90° (dashed and grey curves, respectively). In line with our hypotheses, distance between the patterns already learnt and the task requirement determines both the routes to learning and the amount of interference.

5. Principle of time scale

In the precedent sections, we have put forth two routes to learning, a "bifurcation"-route and a "shift"-route to learning. Looking a bit closer to this issue, it appears that "bifurcation"-route leads to a simultaneous improvement in accuracy and in stability (cf. Figure 9), whereas the "shift"-route entails a loss in stability first, due to the flattening of the corresponding basin of attraction. Now, if the task constraint continues to further impinge on the system dynamics, the basin of attraction gets deeper and narrower. So, accuracy might meet its definitive value, faster than stability.

A main feature of this picture is that the "shift"-route to learning implies a strong discrepancy between accuracy and stability. A required pattern may be well performed in terms of accuracy, while it still lacks stability. Our contention is to set a principle that *the relevant time scale of learning is the time scale of the new pattern stabilization*. Time scale refers here to the temporal resolution and the length of the time period at which significant changes in behavior may be captured. So, the relevant feature of learning is the evolution of stability rather than that of accuracy. A tricky problem, however, is to ponder the extent to which the two routes to learning are actually separate processes.

5.1 Memories distortion

A substantial contribution to the issue of the time scales of accuracy and stability was provided in a study on spatial memory by Giraudo and Pailhous (1999). Participants were presented a target pattern of dots that they had to memorize for five seconds. Then, the target pattern was withdrawn and they had to reproduce it from memory as accurately as possible, twenty times on a raw. On trial 21, participants studied the target pattern once again for five seconds and then 20 additional reproduction trials were administered. The analysis of results was focused on the time course of accuracy and variability. Accuracy was assessed by the discrepancy between the reproduced Figure and the target configuration, and variability by the discrepancy between two consecutive pattern restitutions. Accuracy measures how much the memorized mental image deviates from the target pattern, while variability measures how much stabilized such a memory is.

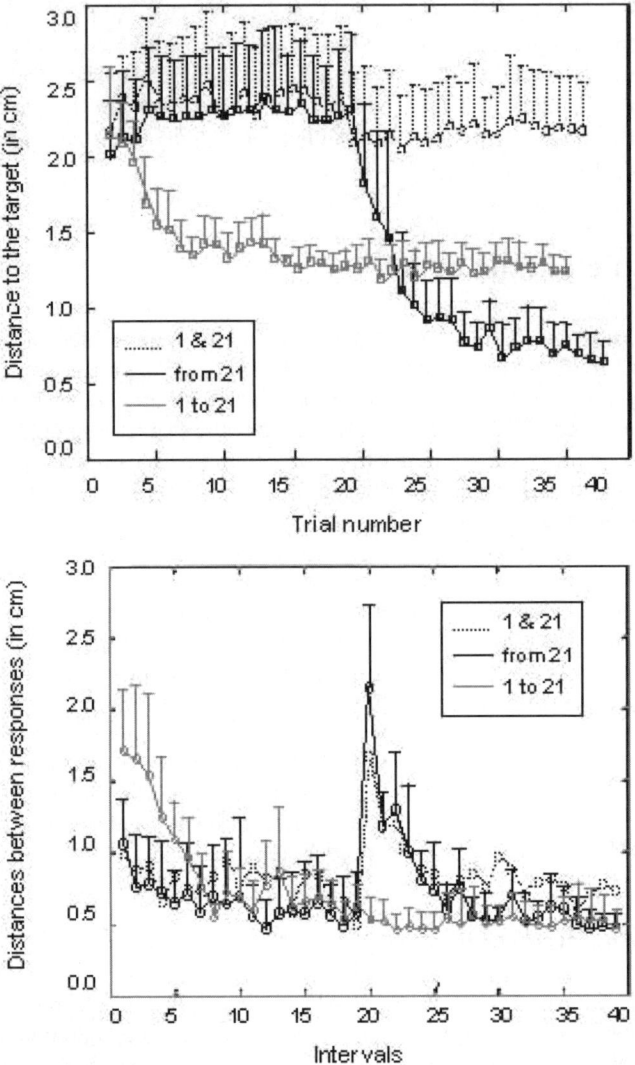

Figure 12. Evolution of response accuracy and variability in a visual dot pattern reproduction.

The evolution of accuracy showed that the memorized configuration first diverged gradually from the target pattern until there was no further improvement over time (top panel in Figure 12). After trial 21, at which the model was displayed once more, an abrupt improvement in accuracy was observed, followed by a progressive leveling-off. Results were quite different for variability (bottom panel in Figure 12). After trial 21, the sudden rise of inter-response variability due to the model presentation was followed by a slow decrease, leading back gradually to a stable memory state. In short, on the last 20 trials, there was an almost immediate stabilization in accuracy, whereas variability still decreased.

These findings suggest that the time scale for accuracy is faster than for variability. The interpretation by the authors was that memory relies on a fast deviation process and a slow structuring process. The structuring process creates links between the points of the target configuration, giving rise to stable clusters. The deviation process shifts the learned pattern away from the target configuration toward the stable clustered configurations. Regarding stability evolution, our interpretation is that loss in stability, signaled by the increase in variability, sets accuracy free to change until the stabilization of a new memory state. This implies that if accuracy is steady while a complete stabilization is not reached yet, it is mistaken to deduce that a stable memory state has been created with learning. Thus, a reliable and exclusive sign of the emergence of a basin of attraction, hence learning, is a decrease in variability, *irrespective of the accuracy attained*. When there is no increase in stability with practice, there is no true learning, but a mere *transient adaptability* to the task requirement. Such a transient adaptability leads back to the issue of the "shift"-route to learning. Our contention is that such adaptability, involving the shift or the deformation of an existing attractor in the direction of the to-be-performed pattern reflects behavioral adaptability on a short time scale. This must be sharply distinguished from *"true" learning*, a slow process leading to a qualitative alteration of the underlying coordination dynamics and implying the creation of an altogether novel stable state, via the "bifurcation"-route.

5.2 Learning and attention

Up to this point, we argued that true learning pertains to the stabilization of an attractor close to the task requirement, a phenomenon empirically signaled by the decrease in accuracy *and* in variability on the same time scale. Recently, several studies looked at another behavioral variable that evolves on the same time scale as stabilization, namely, attentional cost (Temprado et al. 1999; Monno et al. 2000; Pellecchia and Turvey 2001; Temprado et al. 2001; Zanone et al. 2001). It was hypothesized that the deeper a basin of attraction is, the lower the mental load expended to sustain the corresponding coordination pattern, and conversely.

Empirically, attentional cost can be assessed through a dual-task paradigm. This method, classic in experimental psychology, involves two simultaneous concurrent tasks. Basically, a secondary, discrete task is used as a probe to evaluate the attentional cost involved in a primary, continuous task. The lesser attention is needed to preserve or improve performance in the primary task, the more may be allocated to the second task, and inversely. Accordingly, combining a continuous bimanual coordination task with a discrete Reaction Time (RT) task, the above studies established a tight covariation between attentional cost and pattern stability: The most stable in-phase pattern was also the least costly RT-wise, as compared to the anti-phase pattern.

In the perspective of learning studies, the foregoing assumptions imply a dynamic interplay between pattern stabilization and attentional cost. Stabilizing a new pattern with learning, which incurs an attentional cost, should result in an evolving trade-off between stability and RT. As learning proceeds and the pattern to be learned stabilizes progressively, more attention should be available to be allocated to the secondary RT task. RT should then improve *pari passu* with the pattern stabilization due to learning.

Recently, Jouet (2002) explored this hypothesis by comparing the evolution of the trade-off in subjects with bistable coordination dynamics, characterized by 0° and 180° preexisting stable patterns. After a scanning probe, participants were asked to learn a new coordination pattern, namely 90°, in a single task and in a double task. In the single task, they had to perform periodic pronation-supination forearm movements with two customized joysticks in accordance with LEDs displaying the to-be-learned phasing pattern. In the dual task, they were required to do the same task, while responding as fast as possible to randomly presented auditory signals by pressing the trigger buttons on both joysticks with their index fingers. After each trial, KR on accuracy, stability and, if available, on TR was returned to the learners, followed by instructions aimed to improve performance. On the basis of KR, learners in the dual-task were forced to perform at their utmost on both learning and TR tasks. They had to try and attain the same level of accuracy, stability, and RT as that attained in pre-test single tasks, in which they had to perform the 0° and the 180° phasing patterns, and the RT task. The whole experimental design comprised 30 single-task trials interspersed with three blocks of five dual-task trials. At the end of the protocol, a final scanning probe was carried out.

The evolution of all three variables is presented in Figure 13 for "bistable" subjects. The three panels represent accuracy ($\triangle$line), SD ($\diamond$line), and RT ($\blacklozenge$line) in the dual task probes, before, during, and after practice. Results reveals that learning the task requirement led to an increase in accuracy and a decrease in variability, as expected, but also in a significant decrease in RT. Here again, stability and attentional cost co-evolve, in the particular case of stabilization with learning.

Results indicated that subjects, who learnt the task requirement by increasing both accuracy and stability, freed attention from the bimanual task as learning proceeded. Thereby, they could allocate these attentional resources just released to the RT task, so that RT task improved over practice. These findings pertaining to the so-called "bifurcation"-route to learning suggest that in addition to variability,

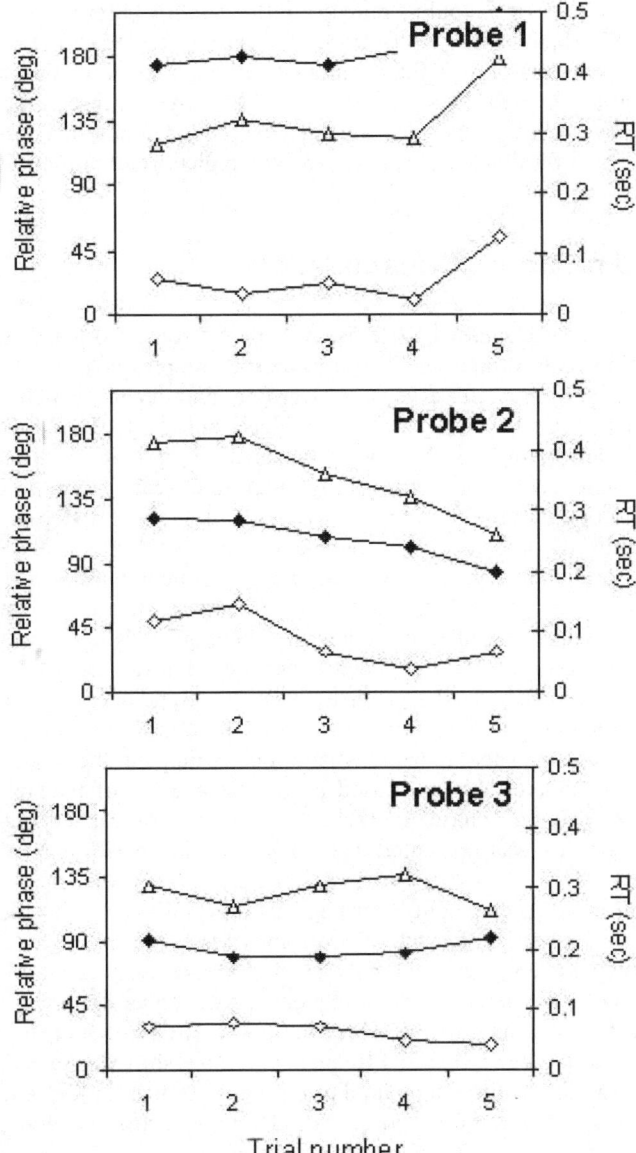

Figure 13. Evolution of accuracy (△), SD (◇), and RT (◆) during learning a 90° bimanual pattern for bistable subjects, at the beginning, in the middle, and at the end of practice (top to bottom panels, respectively).

attentional cost may be a second key clue to capture the differences between true learning and transient adaptability. In the latter case, although performance might be pretty accurate, the attentional cost needed to sustain a still marginally stable coordination pattern should remain rather high. Preliminary results on "multistable" subjects who had to learn 135° atop of already existing 0°, 90°, and 180° stable patterns (ibid.) indicated that although performance accuracy was attained, variability *and* RT were still large. This suggests that in the "shift"-route to learning, an acceptable performance may be produced in terms of accuracy, if in an inappropriate fashion stability-wise and at a high cost attention-wise.

6. New directions: Consolidation

Thus far, we have discussed the existence of two routes to learning and pointed out their time scale properties. These properties might open an avenue to tackle more subtle and intricate effects of learning and recall, which are tricky to interpret in a coherent fashion. A leading idea would be to distinguish the various memories contributing to the total coordination dynamics in terms of their respective time scales. Experimentally, the time scale of a process may be captured through its time of relaxation (cf. Eq. 2.1.1). A first step in this direction was accomplished in a study by Zanone and Athènes (1999) on the dynamics of a single temporal memory. The crux of the experiment was to "pull away" the performance of a cadence from a learned frequency in order to investigate the relaxational properties of the underlying memory. Using a scanning probe requiring to perform 16 frequencies between 0.5 Hz to 2.0 Hz, individual preferences were established so that the least stable and accurate frequency could be set as the learning task. This frequency was practiced over 60 trials, which led to a strong learning. Then, unbeknownst to the participant, the model cadence was gradually and undetectably increased by a total amount of 0.2 Hz. Immediately after such a "driving" phase, a "relaxation" phase was carried out for 75 s, in which no stimulus was presented any longer, so that performance could evolve freely.

The evolution of performance through the driving and relaxation phases is displayed in Figure 14. During the driving phase, in accordance to the task requirements, the error curve progressively veered away from the null value, corresponding to the newly learned frequency. As soon as the pacing stimulus was removed at the beginning of the relaxation phase, the error first shifted swiftly in the direction of the newly learned frequency. Such a shift was expected according to the literature on continuation paradigm, suggesting that performance assessed in the presence of the stimulus does not reflect the actual state of the memory variable. Then, the error returned progressively toward the last frequency displayed and performed in the driving phase.

These findings are puzzling, in that instead of the expected, exponential relaxation toward the learned attractor (overdamped relaxation), the behavioral trajectory took a more complicated path. A possibility is that another type of relaxation (underdamped) that leads to an oscillatory trajectory to the attractor might be considered. Still, it is remarkable that behavioral fluctuations were not centered

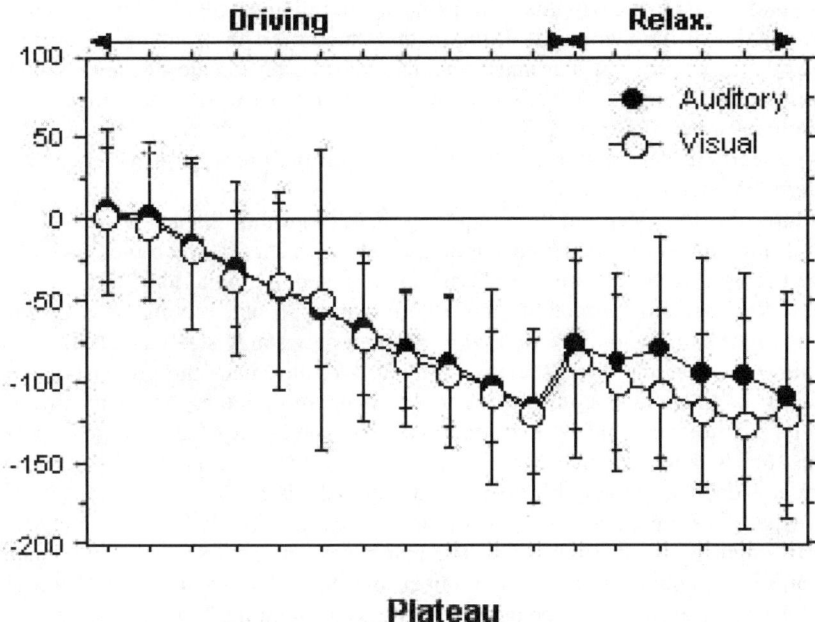

Figure 14. Evolution of the performed frequency in the driving and relaxation phases.

about the learned attractor. This finding is reminiscent of the well-known phenomenon of consolidation, which refers to the delayed and undercover reinforcement of memory over time, thoroughly studied in the literature over the time scales of seconds, hours, and days. An exciting challenge now is to try and interpret these multiple effects in terms of spontaneous changes or exchanges of stability among existing memories, which appear to evolve on different time scales. In the current state of knowledge, however, we prefer to refrain from too farfetched speculations.

7. Conclusion

The goal of this chapter was to promote an idea that, despite the stunning diversity of phenomena occurring in learning systems, generic principles governing the acquisition of new behavior may be sought for. If we believe in some kind of coherence underlying a vast phenomenological diversity, we should benefit from the concepts and tools provided by universal theoretical frameworks, such as self-organization theories and dynamical system models, which may help organizing, enriching, and eventually making sense of the huge bulk of experimental data.

Along these lines, we used an operational approach to discuss three general principles putatively governing learning and memorization. The principle of symmetry specifies that in every symmetrical learning system, stabilization of a new pattern that breaks this symmetry is complemented by the spontaneous stabilization of the symmetrical, unpracticed pattern. The principle of time scale states that time scale of learning coincides with the time scale of the new pattern stabilization, signaled empirically through a decrease in variability and in attentional cost. The principle of distance posits that the evolution of stability during learning depends on the distance between the preexisting patterns and the pattern to be learned.

We turn now to a deep question lingering throughout this chapter, that is, how the initial attractor layout is actually modified with practice. Two routes to learning have been put forth: a "shift"-route, which corresponds to the local alteration of a preexisting basin of attraction, and a "bifurcation"-route, relying on a profound change in the intrinsic dynamic giving rise to a new stable state. The "shift"-route is followed when the distance between the task requirement and preexisting stable patterns is small, and conversely for the bifurcation route. An analysis of the memory distortions and of the attentional cost reveals typical features for both types of dynamic changes, which may be captured through the increase in accuracy, decrease in variability and in attentional cost.

But what does the existence of two routes to learning imply? At first glance, the "shift"-route to learning modifies the parametric properties of the coordination dynamics without changing the number of attractors, so that the topological properties of the system remain invariant. It is tempting to assume that such changes are temporary, at least in comparison to the other route. In contrast, the "bifurcation"-route leads to changes that alter the dynamics qualitatively, modifying thus the topology of the system. Those changes may not be so easily reversible. As discussed by Jagacinski et al. (2000), such a distinction between topological and metrical changes may follow merely from properties of the mathematical tools, without any real, behavioral counterpart in the skill under study. A second possibility is that this distinction corresponds to two distinct levels of behavioral organization distinguished by Bernstein (1967), the properties of which have been captured by the models of schemas (Schmidt and Lee, 1999) and motor programs (Keele et al., 1990). The dynamical details of this distinction, however, remain to be fully developed and tested.

References

Adams JA (1987) Historical review and appraisal of research on the learning, retention, and transfer of human motor skills. Psychol Bull 101, 41-74

Alba JW, Hasher L (1983) Is memory schematic? Psychol Bull 93, 203-231

Anderson JR (2001) The power law as an emergent property. Mem Cognition 29, 1061-1068

Anderson JR, Schooler LJ (1991) Reflections of the environment in memory. Psychol Sci 2, 396-408

Atchy-Dalama P (2000) Spontaneous evolution of coordination dynamics with unspecific practice. Universite Paul Sabatier Toulouse, Unpublished Master Thesis.

Ballard B (1913) Obliviscence and reminiscence. Brit J Psychol 2, 100-254

Bartlett FC (1932) Remembering: A study in experimental and social psychology. Cambridge University Press, London

Bernstein NA (1967) The coordination and regulation of movements. Pergamon, Oxford

Brashers-Krug T, Shadmehr R, Bizzi E (1996) Consolidation in motor memory. Nature 382, 252-255

Buschke H (1974) Spontaneous remembering after recall failure. Science 184, 579-581

Conway MA (1991) In defense of everyday memory. Am Psychol 46, 19-27

Estes WK (1997) Process of memory loss, recovery and distortion. Psychol Rev 104, 148-169

Fontaine RJ, Lee TD, Swinnen S (1997) Learning a new coordination pattern: Reciprocal influences of intrinsic and to-be-learned patterns. Can J Exp Psychol 51, 1-9

Fuchs A, Jirsa VK (2000) The HKB model revisited: How varying the degree of symmetry controls dynamics. Hum Movement Sci 19, 425-449

Giraudo MD, Pailhous J (1999) Dynamic instability of visuospatial images. J Exp Psychol Human 25, 1495-1516

Glenberg AM (1997) What is memory for. Behav Brain Sci 20, 1-55

Haider H, Frensch PA (2002) Why aggregated learning follows the power law of practice when individual learning does not. J Exp Psychol Learn 28, 392-406.

Haken H (1983) Synergetics: An introduction, 3 edn. Springer-Verlag, New York

Haken H (1993) Basic Concepts of Synergetics. Appl Phys A 57, 111-115

Haken H, Kelso JAS, Buntz H (1985) A theoretical model of phase transitions in human hand movements. Biological Cybernetics 51, 347-356

Haken H, Wischert W, Wunderlin A, Meijer G (1993) Introduction of synergetics. In: Greppin H, Bonzon M, Agosti RD (eds.) Some physicochemical and mathematical tools for understanding of living systems. University of Geneva, Geneva, 71-87

Haken H, Wunderlin S, Yigitbasi S (1995) An introduction to synergetics. Open Syst Inf Dyn 3, 97-130

Heathcote A, Brown S, Mewhort DJK (2002) The power law repealed: The case for an exponential law of practice. Psychon B Rev 7, 185-207

Hodges JN, Franks IM (2000) Attention focusing instructions and coordination bias: Implications for learning a novel bimanual task. Hum Movement Sci 19, 843-867

Hodges JN, Franks IM (2001) Learning and coordination skill: Interactive effects of instruction and feedback. Res Q Exercise Sport 72, 132-142

Jagacinski RJ, Peper CE, Beek PJ (2000) Dynamic, stochastic, and topological aspects of polyrhythmic performance. J Motor Behav 32, 323-336

Jouet I (2002) Evolution of attentional cost with learning a non-spontaneous coordination pattern. Universite Paul Sabatier Toulouse, Unpublished Master Thesis.

Keele SW, Cohen A, Ivry RB (1990) Motor programs: Concepts and issue. In: Jeannerod M (ed) Attention and performance XIII. Erlbaum, Hillsdale, NJ, 77-110

Kelso JAS (1984) Phase transitions and critical behavior in human bimanual coordination. Journal of Physiology, Regulatory, Integrative and Comparative Physiology 15, R1000-R1004

Kelso JAS (1995) Dynamic patterns: The self-organization of brain and behavior. The MIT Press, Cambridge

Kelso JAS, Jeka JJ (1992) Symmetry breaking dynamics of human multilimb coordination. J Exp Psychol Human 18, 645-668

Kelso JAS, Zanone PG (2002) Coordination dynamics and transfer across different effector systems. J Exp Psychol Human 28, 776-797

Koffka K (1935) Principles of gestalt psychology. Hartcourt Brace, New York

Koriat A, Goldsmith M (1996) Memory metaphors and the real life / laboratory controversy: Correspondence versus storehouse conceptions of memory. Behav Brain Sci 10, 167-228

Kostrubiec V, Zanone PG (2002) Memory dynamics: Distance between the new task and existing behavioral patterns affects learning and interference in bimanual coordination. Neurosci Lett 331, 193-197

Lee TD, Swinnen S, Verschueren S (1995) Relative phase alterations during bimanual skill acquisition. J Motor Behav 27, 263-274

Loftus EF, Palmeri TJ (1974) Reconstruction of automobile destruction: An example of the interaction between language and memory. J Verb Learn Verb Be 13, 585-589

Logan GD (1988) Toward an instance theory of automatization. Psychol Rev 95, 492-527

Monno A, Chardenon A, Temprado JJ, Zanone PG, Laurent M (2000) Effects of attention on phase transitions between bimanual coordination patterns: A behavioral and cost analysis in humans. Neurosci Lett 283, 93-96

Neisser U (1967) Cognitive Psychology. Appletin Century - Crofts, New York

Neisser U (1981) John Deans's memory: A case study. Cognition 9, 1-22

Newell KM, Liu YT, Mayer-Kress G (2001) Time scales in motor learning and development. Psychol Rev 108, 57-82

Palmeri TJ (1999) Theories of automaticity and the power law of practice. J Exp Psych Learn 25, 543-551

Pavlov I (1932) Reflexes conditionnels et inhibitions. Gonthier, Genève

Pellecchia GL, Turvey MT (2001) Cognitive activity shifts the attractors of bimanual rhythmic coordination. J Motor Behav 33, 9-15

Piaget J, Inhelder B (1976) La psychologie de l'enfant. PUF, Paris

Schacter DL, Norman KA, Koutstaal W (1998) The cognitive neuroscience of constructive memory. Annu Rev Psychol 49, 289-318

Schmidt RC, Lee TD (1999) Motor control and learning. Human Kinetics, Champaign

Schöner G (1989) Learning and recall in a dynamic theory of coordination patterns. Biol Cybern 62, 39-54

Schöner G, Dijkstra TMH, Jeka JJ (1998) Action-perception patterns emerge from coupling and adaptation. Ecol Psychol 10, 323-346

Schöner G, Haken H, Kelso JAS (1986) A stochastic theory of phase transitions in human hand movement. Biol Cybern 53, 247-257

Schöner G, Kelso JAS (1988a) Dynamic pattern generation in behavioral and neural systems. Science 239, 1513-1520

Schöner G, Kelso JAS (1988b) A dynamic pattern theory of behavioral change. J Theor Biol 135, 501-524

Schöner G, Kelso JAS (1988c) A synergetic theory of environmentally -specified and learned patterns of movement coordination. I. Relative phase dynamics. Biol Cybern 58, 71-80

Schöner G, Zanone PG, Kelso JAS (1992) Learning as change of coordination dynamics: Theory and experiment. J Motor Behav 24, 29-48

Shadmehr R, Holcomb H (1997) Neural correlates of motor memory consolidation. Science 277, 821-825

Shaw RE, McIntyre M, Mace W (1974) The Role of Symmetry in Event Perception. In: Macleod RB, Pick HL, (eds.) Perception: Essays in Honor of James Gibson. Cornell University Press, Ithaca, 650-660

Smethurst CJ, Carson RG (2001) The acquisition of movement skills: Practice enhances the dynamic stability of bimanual coordination. Hum Movement Sci 20, 499-529

Temprado JJ, Zanone PG, Monno A, Laurent M (1999) Attentional load associated with performing and stabilizing preferred patterns. J Exp Psychol Human 25, 1579-1594

Temprado JJ, Zanone PG, Monno A, Laurent M (2001) A dynamical framework to understand performance trade-off and interference in dual tasks. J Exp Psychol Human 27, 1303-1313

Thompson DAW (1961) On Growth and form. Cambridge University Press, Cambridge

Turvey MT, Kugler PN (1984) An ecological approach to perception and action. In: Whiting HTA (ed.) Human motor actions, Bernstein Reassessed, Vol 17. Elsevier Publishers, BV, North-Holland, 373-412

Turvey MT, Shaw BK (1995) Toward an ecological physics and a physical psychology. In: Solso RL, Massaro DW (eds.) The science of the mind: 2001 and beyond. Oxford University Press, New York, 144-169

Turvey MT, Shaw BK, Reed ES (1981) Ecological laws of perceiving and acting: In reply to Fodor and Pylyshyn (1981). Cognition 9, 237-304

Ward LB (1937) Reminiscence and learning. Psychol Monogr 44, 200-245

Wenderoth N, Bock O (2001) Learning of a new bimanual coordination pattern is governed by three distinct processes. Motor Control 1, 23-35

Wenderoth N, Bock O, Krohn R (2002) Learning of a new bimanual coordination pattern is influenced by existing attractors. Motor Control 6, 166-182

Zanone PG, Athènes S (1999) Frequency learning as the strengthening of a memorized attractor. In: 10th International Conference on Perception and Action. Edinburgh

Zanone PG, Kelso JAS (1992) Evolution of behavioral attractors with learning: Nonequilibrum phase transitions. J Exp Psychol Human 18, 403-421

Zanone PG, Kelso JAS (1994) The coordination dynamics of learning: Theoretical structure and experimental agenda. In: Swinnen S, Heuer M, Massion J, Casaer P (eds.) Interlimb coordination: Neural, dynamical and cognitive constraints. Academic Press, New York, 571-593

Zanone PG, Kelso JAS (1997) Coordination dynamics of learning and transfer: Collective and component levels. J Exp Psychol Human 23, 1454-1480

Zanone PG, Monno A, Temprado JJ, Laurent M (2001) Shared dynamics of attentional cost and pattern stability. Hum Movement Sci 20, 765-789

PART III: COORDINATION DYNAMICS OF POSTURE: CONTROL MECHANISMS

Using Visual Information in Functional Stabilization: Pole-Balancing Example

Gonzalo C. de Guzman

Center for Complex Systems and Brain Sciences, Florida Atlantic University, Boca Raton FL 33431

1. Introduction

Some of the common tasks that confront us in our day-to-day living involve stabilizing unstable situations. Typical examples are maintaining posture (Jeka & Lackner, 1995), or trying to stand still or preventing falls when walking. Coordinating articulator movements, if the required coordination pattern is not in our repertoire of stable movement activities, can also be considered as a stabilization problem (Zanone & Kelso, 1992). And of course, there is the problem continuously balancing an object that will not balance on its own. The last problem has traditionally been addressed with considerable success within the domain of physical control theory (e.g., see Kwakernaak & Sivan, 1972). Our focus is on human pole balancing and the relevant information involved accomplishing the task.

In balancing tasks, relevant information is used by humans to stabilize the unstable system (Kelso, 1998). Such information may come in varied forms and modalities. Posture can be stabilized by light haptic contact of finger with a touchbar (Jeka & Lackner, 1994; Jeka, Schöner, Dijkstra, Ribeiro, & Lackner, 1997). That posture can also be affected by changes in optic flow across a range of ages and motor developmental stages is demonstrated in the moving room paradigm (Berthenthal & Bai, 1989; Berthenthal, Rose, & Bai, 1997; Lee & Aronson, 1974). In the stabilization task, the questions we pose are the following: (1) what information is readily available to a human controller; and (2) how may one use this information in a human control system? When balancing the pole with the hand, we typically use both the visual and haptic senses. If one looks away momentarily, it is still possible to do the task if the person is skilled enough; suggesting that haptic information alone is sufficient in some situations. For the average person, it is most likely that one guides hand movements visually, especially during critical situations. Visually, a pole balancer is presented with an image of where the pole is as well as how fast it is moving towards or away from the vertical. Although these are readily available data, they may not be the best format for conveying the relevant and immediate information for instantaneous control of the pole.

Here, it is proposed that more readily assimilated information is the combination

$$\tau_{bal} \equiv \frac{\theta}{\dot{\theta}} \qquad (1)$$

which we refer to as the time-to-balance. Here, θ is the vertical pole angle, and $\dot{\theta}$ its instantaneous velocity. Although, this variable imparts partial position and velocity information only, it is perceptually direct and more efficient carrier of the relevant information for the task. When the pole is moving towards the vertical, the time-to-balance is the time it takes for the pole to reach the vertical from its current position if the current velocity is maintained. The definition is motivated by a similar variable used in the study of optic flow, which in the current context basically says the following: if an extended object is looming towards you at a constant velocity, its projection, A, on the retina expands correspondingly in such a way that the inverse of the relative rate of expansion (also referred to as the optic-τ) is the time-to-contact with this oncoming object (see e.g., Lee, 1976).

$$\tau \equiv \frac{A}{\dot{A}} \qquad (2)$$

When the size of the oncoming object is manipulated to induce a false estimate of the collision time, results showed reactions based on the perceived (as measured in τ) rather than the actual time-to-contact (Savelsbergh, Whiting, & Bootsma, 1991; Savelsbergh, 1995).

Although the concept is closely linked to optic flow studies, we believe that the time-to-balance information has its own merit. For example, it is locally robust under some realistic transformation of the visual information. The simple scaling of the pole angle, $\theta \to \lambda\theta$, produces the same estimate of the time-to-balance. This may be realized, for example, when the balancer moves around the pole so that the perceived angle is smaller than it actually is. The transformation $\theta \to L \cdot \sin \theta$, where L is the pole length also preserves the time-to-balance, at least for small angles, and may be used when monitoring how far the tip of the pole is from the vertical. It is clear that any transformation $z \equiv T(\theta)$, with $T(0) = 0$ and $T'(0) \neq 0$, preserves time-to-balance information when around the balance point. A key point is that if indeed the time-to-balance information based on the pole angle is being utilized, a transformation of the visual field that yields another kinematic variable z that, locally and near the corresponding balance point z = 0, behaves like above also is available for estimating the time-to-balance. Our approach treats functional stabilization as a collision problem with respect to the balance point on a cycle-to-cycle time scale.

2. Pole-balancing experiment

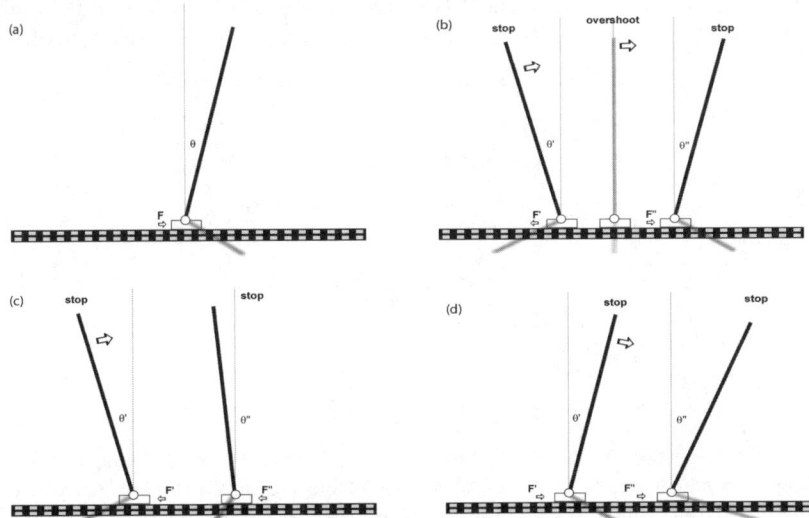

Figure 1. (a) Setup for human pole balancing experiment. To control the pole, subjects hold and apply a force F to a base constrained to move along a linear track. Pole motion is limited to the vertical plane and swings around a pivot point at the base. Between pole stops, only three non-catastrophic behaviors are possible: (b) a pole crossover; (c) an undershoot; and (d) a drift. Consequently, there are only nine contiguous two-segment behaviors and they correspond to all possible pairings of (b), (c), and (d).

In Figure 1a, we see the basis configuration (based on the design by Barto, Sutton, and Anderson (1983)) used in our pole-balancing experiment. The components are: pole (108 cm long, mass of 207.4 g), horizontal track (1.8 m long). and base cart (mass of 388 g). Movement data of hand and pole were collected from five infrared sensors strategically located to yield pole angle and hand position sampled at 100 Hz. More details of the set up can be found in Foo, Kelso, and de Guzman (2000). The task was to balance the pole by moving the cart with right hand along the linear track (the x-axis). Haptic information from the cart, but not directly from the pole itself, was thus available to the participant. The goal was to balance the pole for 30 sec without allowing it to fall and make contact with the track. Failure to complete 30 sec of continuous balancing constituted an unsuccessful trial.

3. Cycle-by-cycle analysis of the time series

Figures 1b,c,d show the three possible movement segments between pole stops excluding catastrophic failures (defined as pole touching the horizontal or the base hitting the track boundaries). In all three figures, the left stop precedes the right stop position in time. A sequence of successful movements is a concatenation of a crossover (Figure 1b), an undershoot (Figure 1c), and a drift (Figure 1d) in any order. For later reference, we label these segments by C, U, and D respectively.

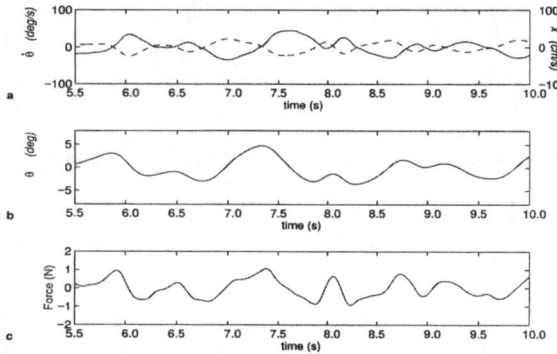

Figure 2. Representative time series showing typical behavior of the pole and hand movement when the pole is close to the vertical. Note the tight coupling between the velocities (a) and the similar shapes of the angle and force (b, c) suggesting a control force that is proportional to the angle. [From Foo et al., 2002]

In Figure 2, we see a representative time series when the pole is hovering near the vertical. Note the tight coupling between hand and pole velocities in Figure 2a. This is partly due to the mechanical connection of the base to the pole. To examine coordinated behavior beyond the mechanical coupling, a cross-correlation was performed between the time series of the hand and pole velocities for each cycle. A measure Δ, defined as the time difference between the peak of the normalized cross-correlation and a zero-lag value (when hand and pole velocity peaks coincide) was used as basis data to examine coupling. The mean value of 10 ms (hand lagging the pole) found during successful balancing suggests a high degree of coupling between the hand and the pole. The force F (plotted in Figure 2c) exerted by the hand on the base was computed numerically from the physical equations of motion using measured positions and size and mass parameters. Comparing Figures 2b and 2c, we note the similarity. This suggests a force proportional to pole angle. How that proportionality "constant" is determined is an important question and as we will show, can be obtained during critical situations from the time-to-balance variables. Note that for strict angle-proportionate control (i.e. $F = k\theta$, k=constant), model simulations near the vertical yield regular harmonic oscillations of the pole. The anharmonicity in the pole time series suggest a more complex behavior on k.

In Figure 3, we show a shorter time series of the velocities and the posited perceptual variables $\tau_{bal}, \dot{\tau}_{bal}$ from a successful trial. We use the convention that negative (positive) pole angles indicate the pole is to the left (right) of the vertical. This has the consequence that negative τ_{bal} means a pole moving up towards the vertical while a positive one indicates a pole moving down away from the vertical regardless of whether the pole is to the left or right of the balance point. Note in Figures 3b how the plots quickly rise up to infinity in absolute value. This is due to the angular velocity going to zero and provides a natural partitioning of the time series. The boundaries correspond to the pole stops shown in Figures 1b,c, d. In the τ_{bal} time series (Figure 3b), a successful balancing segment is a sequence of three basic curves: a sigmoid, a U-shape, and an inverted U-shape. The sigmoid corresponds to a crossover of the vertical; starting from an infinitesimal velocity going up towards the vertical (yielding infinitely negative τ_{bal}) and ending up on the other side and stopping (for the infinitely positive τ_{bal}). An inverted-U curve means the pole starts from stop, moves up, then stops but fails to cross over the vertical (an undershoot). In a normal-U curve segment, the pole starts from a stop, falls away from the vertical, and then stops. This corresponds to a drift and occurs on the way down.

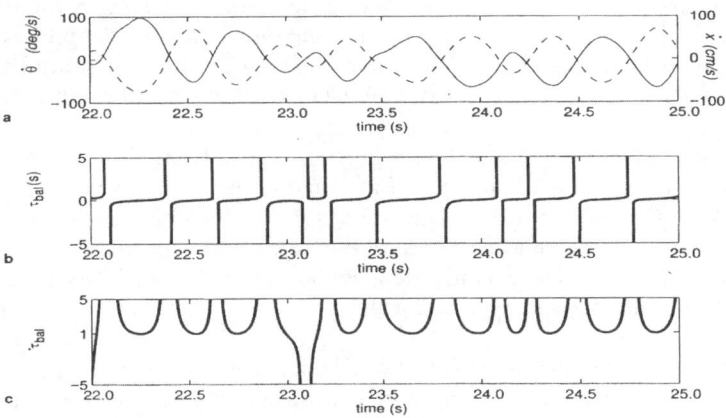

Figure 3. Time series of the (a) velocities, and (b,c) the posited perceptual variables. Negative (positive) values in (b) correspond to pole moving towards (away) from the vertical. Near the pole stops, the time-to-balance variable goes to $\pm \infty$. The sigmoid, U-

shape, and the inverted U-shape curves in (b) correspond to crossovers, drifts, and undershoot, respectively. [From Foo et al., 2002]

To explore action-perception coupling that facilitates successful balancing, we examine first how failure typically occurs and what the corresponding successful potential recovery would be. By definition, a failure occurs when the second stops in Figures 1b,c, d do not materialize before the pole hits the horizontal or before the subject intervene by catching it. A failure can occur during a crossover or a drift. On the other hand, an undershoot can transform into a failure only if it overshoots the vertical and thus become a failure from a crossover. There are therefore only two possibilities for the end cycle of a catastrophic run, an incomplete crossover and an incomplete drift. A close examination of all the time series show failure predominantly due to an incomplete drift. Since successful balancing also involves anticipation, we now look at what happens during a cycle immediately preceding a catastrophic failure (an incomplete drift). First, we define a path to be any two-cycle sequence in a successful balancing. There are nine possible paths and these correspond to all paired combination of the basic movements in Figures 1b,c,d. Empirically, what is observed is only a smaller subset {CC, CD, CU, UD, DC, DU} where C, D, and U refer to cross-over, drift, and undershoot, respectively. Since a failure predominantly end in an incomplete drift, we describe the paths CD and UD as potential paths to failure.

What transpired when the participant allowed to pole to fall after a crossover movement (analogous to the path CD in the successful balancing). After a successful crossover, the hand decelerates such that the pole velocity approach zero. The pole had not reversed directions, however, and the pole moved away from the vertical in the same direction it crossed it. In order to continue with successful balancing, the participant had to accelerate the pole in the direction opposite the pole's fall and also in the opposite direction from the previous hand movement. If the participant failed to make this successful reversal of the hand, the pole falls catastrophically. Thus, path CD *potentially* could result in a *failure to reverse*.

Consider now the case when failure follows an undershoot. Here, the participant undershot the vertical and then moved the hand in the direction of the pole's fall enough to decelerate the pole to near stand still. As the pole falls the participant did not make a successful acceleration of the hand in the same direction as the fall of the pole (and the current direction of the hand). We call this a *failure to continue* and path UD is its *potential* path.

In a successful execution of paths CD or UD, it is clear that the participant had to perform an active intervention (more so than when going via the other four paths) to prevent a failure. We may thus view paths CD and UD critical situations since these are instances of successfully preventing a failure to reverse and failure to continue, respectively. It is postulated that when executing these paths participants must pay special to the relevant information in order to sustain successful balancing. It is especially during these situations that we chose to examine the relation between perceptual variables and pole motions in greater detail (see Savelsbergh, Whiting, & Bootsma (1991) for a similar approach).

4. Correlation studies

What evidence do we have that the time-to-balance information is being utilized in pole-balancing task? Since it is possible that the controller may not actually be looking at the pole all the time, hand action may not always be tightly correlated to visual pole behavior. We therefore look at points of 'interceptive action' (Warren, 1988) defined here as the time points of extremum velocity, or equivalently, points of zero hand acceleration,

$$\ddot{x} = 0 \tag{3}$$

These are points where the hand starts to reverse itself in anticipation of a future corrective action. Thus, there is a delay between the initiation of the intent to reverse and the actual reversal of the pole. Note that reversal per se does not signify attentive anticipation. To see this, one may wiggle the pole at a fast frequency and keep it near vertical for some time. This involves a lot of reversals yet does not require full anticipation of the outcome. Ultimately, such a strategy tends to fail because the action does not match the intrinsic temporal characteristic of the pole. Just before a potential failure, and to correctly balance the pole, the reversal point is adjusted (delayed or advanced) so that succeeding cycles again become manageable. More specifically, we want to see which pole variables, evaluated at the interceptive points are predictive of the half-period of the hand velocity. Put another way, we want to determine which pole measure (evaluated at a single time point) consistently predicts the (longer in time-scale) half-period of hand velocity.

For cycle-to-cycle correlation analysis, we used the movement partitions bounded by pole stops (see Figures 1b,c,d). Values of the posited perceptual variables $\theta, \dot{\theta}, \dot{x}, \tau_{bal}, \dot{\tau}_{bal}$ at time points of peak hand velocity were then correlated with the half-period of the corresponding hand velocity oscillation cycle. The underlying assumption is that the half-period of hand action is an approximation to the "time to upright" the pole and is modulated directly by the perceptual information. The correlation coefficients reported in Table 1 are partially reproduced from Foo et al. (2000) and includes result from analysis of 36 subjects (18 males, 18 females) balancing both straight and L-shaped poles. The values in column 2 come from analysis on kinematic data collapsed across all cycles. No significant relationships were found. The six empirical two-cycle paths (CC, CD, CU, UD, DC, DU) defined previously label the last six columns. Note the highly significant inverse correlation ($r = -.96$) found between τ_{bal} and $\dot{x}$ period in path CD (crossover to drift). No similar strength relationship between τ_{bal} and $\dot{x}$ was found in path UD (undershoot to drift). The presence of the significant τ_{bal} and $\dot{x}$ suggests that participants may be sensitive to the perceptual variable τ_{bal} especially during the critical situation of avoiding a failure to reverse. Note that our data do not show a relationship between τ_{bal} and $\dot{x}$ in non-critical paths during successful performance.

Perceptual variable	All cycles	Categorized by τ_{bal} path					
		CC	CD	CU	UD	DC	DU
τ_{bal}	-.01	-.33	-.96	-.40	.13	-.03	-.04
$\dot{\tau}_{bal}$	.09	-.06	.38	-.18	.26	-.14	-.08
$\dot{x}$	.04	.06	.14	.07	.08	-.02	-.02
θ	.03	.10	.19	.02	.13	-.01	-.06
$\dot{\theta}$	-.04	-.05	-.11	-.07	-.07	.01	-.01

Table 2. Correlations Between Hand Motions and Perceptual Variables.

5. Using perceptual information for balancing

Here, I present a brief outline of how one may use this perceptual information to control the system. Many things are left out, more specifically the translational dynamics of the hand motion, and we concentrate only on the behavior of the pole angle (see Foo, 2000 for a full analysis). If the base of the pole fixed and the pole is close to the vertical, the pole will fall with a characteristic time that depends on the pole's mass distribution. Thus one may have approximately, $\ddot{\theta} = k'\theta$, where k' is a constant that depends on the mass distribution (more accurately, on the radius of gyration about the pivot point of the pole). Assume now that the hand acts in such a way that the force F on the base is proportional to the pole angle. For motion near the vertical, so that both the angle and angular velocity are small, linear analysis shows that the equation of motion for the angle is still specified by the above differential equation except that now the constant k' is a function of both the mass (and its distribution) and the external force constant (see e.g., Ogata, 1978). If the hand is now allowed to move the base of the pole to control it, it is reasonable to assume an equation of motion of the form $\ddot{\theta} = k\theta$ where now k is a time varying function, reflecting the collective effect of physical parameters as well as perceptual information. This can be recast into the form

$$k = \frac{\dot{\tau}_{bal} - 1}{\tau_{bal}^2} \tag{4}$$

where τ_{bal} is the instantaneous time-to-balance defined before. The key point here is that by gauging the time-to-balance variables, one can adjust the k, depending on what might be a suitable strategy for the collision. A strategy might be to adjust the k so that during an approach to the balance point the pole reaches the vertical with a zero velocity (a soft contact). This can be achieved by incrementally adjusting k so that for the current perceived time-to-balance, the rate of change of the time-to-balance is approaches 0.5 (see Lee, 1976).

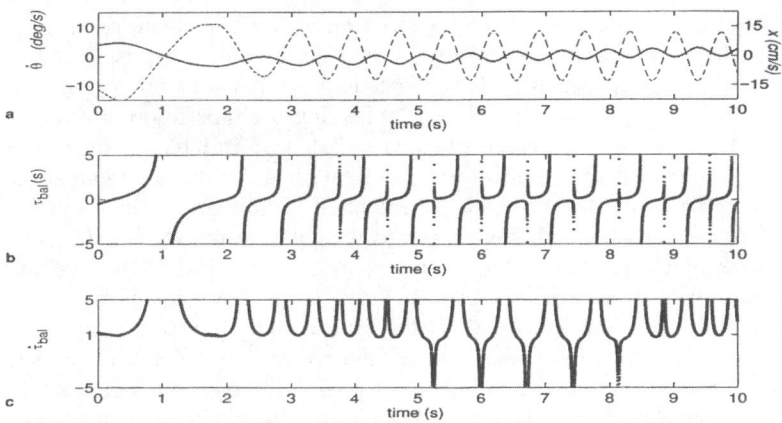

Figure 4. Simulation of pole balancing using the time-to-balance variable and its derivative as the basic perceptual input information used to modulate the proportionality "constant' k. The computation uses the experimental mass and length parameters of the basis apparatus.

In Figure 4, we show a simulation from the physical equations of motion of the cart-pole system (from Foo et al., 2000) using 'proportional' control but where the proportionalityconstant is replaced by time varying function whose value is incrementallyadjusted depending on the input time-to-balance variables. Note that this approach has a slight element of an anticipatory dynamics since the parameter adjustment is based on the perceived time-to-balance. In a recent study, Cabrera & Milton (2002) showed that the small movements one makes when balancing a stick occur at a time scale faster (10 ms) than our typical response time (100 ms). The implication is that stick balancing is more of a stochastic and intermittent process and less of a controlled deterministic one. For example, they allowed the possibility that some of the movements the human balancer makes are beyond his control. They also propose that noise in the nervous system helps in the stabilization process. Our behavioral data differs from those of Cabrera & Milton (2002) because of the differences in size and mass parameters as well as the experimental constraints (pole balancing on a linear track versus stick balancing on a finger). A key difference though is that whereas we focused on the perceptual information during pole balancing, they explored the basic dynamics of the balancing task using a physical approach.

6. Conclusion

In pole balancing, several sources of information are available to gauge the current 'state' of the pole, the human controller, as well as the 'state' of the controller 'in relation' to the pole. The ones that easily come to mind are the pole and hand phase variables (configuration variables), the pressure on the hand, and the stance of the controller, to name a few. In such a complex task, on must be able to process available information quickly. Focusing on the relevant information that efficiently describes the state of the pole as well as the immediate result of ones actions facilitates successful balancing. Using correlation analysis, we showed that during critical situations when there is potential for failure, the hand movement period (which can be taken as reflecting the time-to-upright the pole) correlates well with the perceived time-to-balance variable. In a model simulation, it was also shown how the task might be accomplished in an artificial linear controller by using visual perceptual information to adjust the proportionality "constant". This is similar to weight adjustment such as found in neuromorphic linear control system. Using actual movement data from balancing experiments, neural network controllers can learn how to mimic humans accurately (Guez & Selinsky, 1988). Using time varying digitized images of the cart pole system as visual inputs, Tolat and Widrow (1988) used a pattern recognizing system that is able to balance a pole. Typically, the weights on this artificial neuromorphic controllers are adjusted based some integrated measure of error. Our update rule is not based on an accumulated error measure but on instantaneous perceived deviations from a required behavior of the time-to-balance.

Acknowledgment

This research was supported by National Science Foundation Grant SBR 9511360, National Institute of Mental Health (NIMH) Grants MH42900 and K05MH01386, and NIMH Training Grant MH19116. All experiments and analyses were conducted in collaboration with Dr. Patrick Foo (Brown University) and Prof. J. A. Scott Kelso (Center for Complex Systems and Brain Sciences at FAU).

References

Barto AG, Sutton RS, Anderson CW (1983) Neuron-like adaptive elements that can solve difficult learning control problems. IEEE T Syst Man Cyb 5, 834-846

Berthenthal BI, Bai DL (1989) Infants' sensitivity to optical flow for controlling posture. Dev Psychol 25, 936-945

Berthenthal BI, Rose JL, Bai DL (1997) Perception-action coupling in the development of visual control of posture. J Exp Psychol Human 23, 1631-1643

Bootsma RJ, Oudejans RRD (1993) Visual information about time to collision between two objects. J Exp Psychol Human 19, 1041-1052

Cabrera JL, Milton JG (2002) On-Off Intermittency in a Human Balancing Task. Phys Rev Lett 89, 158-702

Foo P, Kelso JAS, de Guzman GC (2000) Functional Stabilization of Unstable Fixed Points: Human Pole Balancing Using Time-To-Balance Information. J Exp Psychol Human 26, 1281-1297

Jeka JJ, Lackner JR (1994) Fingertip contact influences human postural control. Exp Brain Res 100, 495-502

Jeka JJ, Schoner G, Dijkstra T, Ribeiro P, Lackner JR (1997) Coupling of fingertip somatosensory information to head and body sway. Exp Brain Res 113, 475-483

Kelso JAS (1998) From Bernstein's physiology of activity to coordination dynamics. In: Latash ML (ed.) Progress in Motor Control. Human Kinetics, Champaign, IL, 203-219

Kwakernaak H, Sivan R (1972) Linear optimal control systems. Wiley, New York

Lee DN (1976) A theory of visual control of braking based on information about time-to-collision. Perception, 5, 437-459

Ogata K (1978) System Dynamics. Englewood Cliffs, Prentice Hall, NJ

Savelsbergh GJP, Whiting HTA, Bootsma RJ (1991) Grasping tau. J Exp Psychol Human 17, 315-322

Treffner PJ, Kelso JAS (1995) Functional stabilization of unstable fixed points. In Bardy BG, Bootsma R, Guiard Y (eds.) Studies in Perception and Action III. Erlbaum, Hillsdale, NJ, 83-86

Tolat VV, Widrow B (1988) An adaptive "broom balancer" with visual inputs. Proceedings of the International Conference on Neural Networks, July, San Diego, CA, II-641-II-647

Zanone PG, Kelso JAS (1992) Evolution of behavioral attractors with learning: Non-equilibrium phase transitions. J Exp Psychol Human 18, 403-421

Wagner H (1982) Flow-field variables trigger landing in flies. Nature, 297, 147-148

Warren WH, Jr (1988) The visual guidance of action. In: Meijer OG, Roth K (eds.) Complex Movement Behavior. Amsterdam, North-Holland, 339-380

Postural Coordination Dynamics in Standing Humans

Benoît G. Bardy

Research Center in Sport Sciences, University of Paris Sud XI, Orsay, France

Human stance requires the coordination of multiple joints. This article examines the dynamics of postural coordination modes involving the torso and legs in the control of stance and stance-related activities. Based on data obtained in various experiments using the same postural tracking task, we provide evidence that postural modes (i) emerge out of the coalescence of multiple constraints, (ii) exhibit persistence and changes that are characteristic of self-organized systems, (iii) are modulated by the actor's intention, and (iv) can be learned by modifying the intrinsic dynamics of the postural system. Similarities between postural phase transitions in humans or non-biological phenomena suggest the existence of general and common principles governing pattern formation and flexibility in complex systems, and circumscribe the generality of neurophysiologically-based theories of postural behavior.

1. Postural coordination dynamics in standing humans

One of the major problems facing movement scientists is how humans and other animals coordinate the multitude of degrees of freedom of their bodies, constraining them to act as a single unit in accomplishing behavioral tasks. Standing, walking, reaching, or hitting a moving object are prosaic examples in which successful performance is based upon, and severely constrained by, a set of neuro-muscular synergies temporally and functionally assembled for the purpose of the task. In less than two decades, the field of coordination dynamics has demonstrated its robustness as a theoretical framework for posing (at least), and solving (at best) the problem of human coordination. Experimental and theoretical studies have provided evidences for the existence of non linear dynamical principles in the domain of human inter-limb coordination (e.g., Carson et al., 1995; Haken et al., 1985; Kelso, 1984; Schöner et al., 1986), intra-limb coordination (e.g., Buchanan et al., 1997; Diedrich and Warren, 1995; Kelso et al., 1991), as well as for coordination of limbs with external events (e.g., Dijkstra et al., 1994; Jeka et al., 1997; Schmidt et al., 1990). These studies have concentrated on coordination of individual segments, such as fingers, arms, legs, or head. Surprisingly, virtually no research in pattern dynamics has concentrated on the problem of postural coordination, i.e., coordination between various body parts that underlies our supra-postural behaviors (see however Saltzman and Kelso, 1985, and Woollacott and Jensen, 1996 for early predictions). In our daily actions such as standing, walking, running or dancing, the very great dimensionality of the body (e. g., some 10^3 muscles and 10^2 joints for humans) needs to be reduced to a controllable system exhibiting order, that is, stable and flexible postural patterns. In posture research, a great number of neurophysiological and biomechanical studies have evidenced the role of local constraints — central command signals or forces — in organizing patterns of whole-body coordination (e.g., Allum et al., 1993; Corna et al., 1999; Horak and Nashner, 1986, McCollum and Leen, 1989; Nashner and McCollum, 1985; Nashner et al., 1989; Paï and Patton, 1997). We argue, however, that these local constraints operate in the context of general principles governing postural pattern formation that remain largely unknown. The present contribution aims at underlying the persistences and changes in human posture that witness these general principles. On the basis of recent experimental work, we provide evidences for similarities between postural coordination in humans and other well-known biological phenomena, suggesting the existence of general and common principles governing pattern formation and flexibility in complex systems. We believe that such an approach circumscribes the generality of neurophysiologically-based or biomechanically-based theories of posture, and can therefore be helpful for building a general theory of postural coordination and control.

2. The postural tracking task: A paradigm for investigating postural pattern formation

We have begun to evaluate the multi-segment postural system as a dynamical system, examining modes of coordination that may exist in standing participants between rotations at the ankles and hips. We have focused on the emergence of postural coordination modes that underlie a supra-postural tracking task, as well as on the constraints that shape coordination dynamics (Bardy and Marin, 1997; Bardy et al., 1999; Marin, Bardy, et al., 1999, Marin, Bardy and Bootsma, 1999). In this series of studies, participants in comfortable bi-pedal stance were instructed to maintain a constant distance and phase between their head and a visual target that oscillated along the line of sight. "Tracking" the target with the head was the specific instruction given to the participants. We measured posture during task performance, and found that participants exhibited only two preferred coordination modes between the ankles and the hips (see Figure 1 for an example). Because of its rhythmical nature (e.g., Yoneda and Tokumasu, 1986), postural coordination patterns could be captured by the collective variable ϕ_{rel}, i.e., the relative phase between ankle motion and hip motion. Two values of ϕ_{rel} consistently emerged: An *in-phase* mode, with the two joints moving simultaneously in the same direction (ϕ_{rel} close to 20°), and an *anti-phase* mode with the two joints oscillating simultaneously in opposite directions (ϕ_{rel} close to 180°).

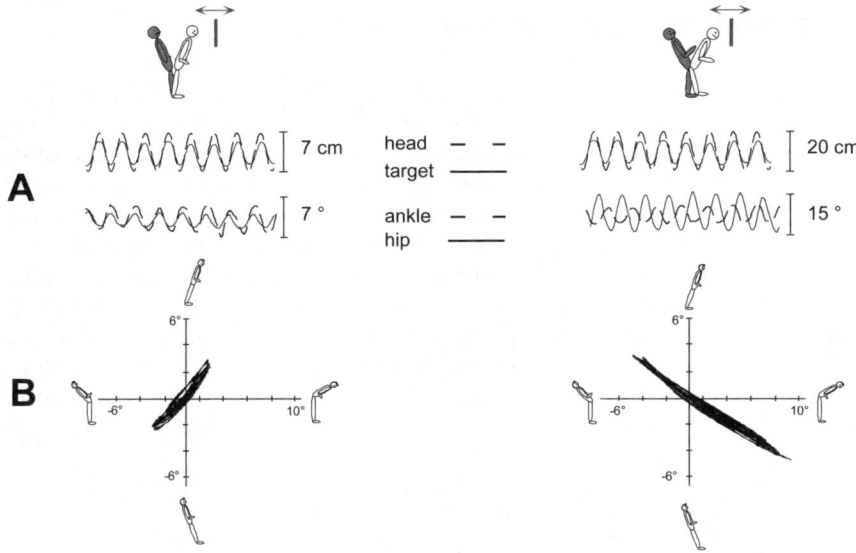

Figure 1. The postural tracking task exhibiting in-phase (left) and anti-phase (right) hip-ankle coordination. (A): Time series of fore-aft positions of head and target showing phase locking of the head (target amplitudes are 5 cm (left) and 18 cm (right)), and related in-phase and anti-phase motion of ankles and hips. (B): Corresponding hip-ankle plane. Figure adapted from Bardy et al. (1999).

From a kinematic point of view, the anti-phase mode corresponds roughly to what Nashner and McCollum (1985) have interpreted as a *hip strategy*. However, the in-phase mode does not have a clear equivalent in the neuromuscular approach developed by Nashner and McCollum. This is because the existence of hip rotation outside of the hip strategy, when it has been acknowledged at all, has not been considered to be a defining feature of multi-segment coordination. As we shall see in the following paragraphs, these two postural modes can be viewed as stable patterns of coordinated movements of the various segments of the postural system. Under the pressure of non-linearities, these patterns emerge via a process of self-organization, resulting in a limited number of stable states. Different, interacting, constraints influence the role played by the non-linearities and thus contribute to the emergence of these patterns, and to changes between them (e.g., Haken et al., 1985; Newell, 1985). Applying the coordination dynamics research agenda to the study of the postural system, we now precise (i) the conditions under which these patterns spontaneously emerge, (ii) bifurcate from one to the other, (iii) are modulated by the actor's intention, and (iv) are destabilized by learning a new, non spontaneous, postural pattern.

3. Constraints on postural coordination

Experimental evidences concerning the role played by a coalition of constraints on the selective emergence of postural states have been obtained. In studies involving the postural tracking task, we have shown that postural modes emerged from the interaction of body-based, task-based, and environment-based constraints (cf. Newell, 1985; Riccio and Stoffregen, 1988). By manipulating target amplitude or frequency, as well as the properties of the body (height of the center of mass, length of the feet, global stiffness) and the surface of support (soft, firm, rolling), evidence was provided that performance at the tracking task was coupled with the mechanical requirement of maintaining the center of mass over the feet (Bardy et al., 1999), and that the form of this coupling depended on the interaction with the surface properties (Marin, Bardy et al., 1999) and the overall stiffness of the body (Marin, Bardy and Bootsma, 1999).

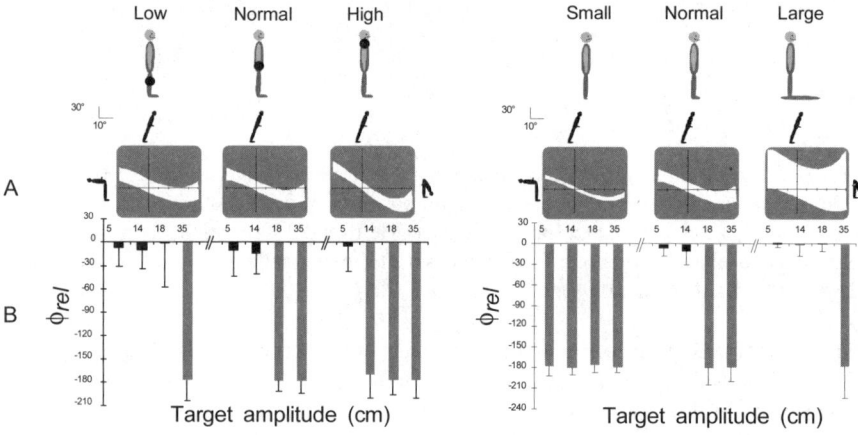

Figure 2. Effects of changes in the location of the center of mass (left), and the length of the feet (right) on (A) the hip-ankle antero-posterior state space, and (B) the hip-ankle relative phase ϕ_{rel}, as target amplitude increases (5, 14, 18, 35 cm). For (A), vertical and horizontal axles represent angular positions of the ankles and the hips, respectively. Angular positions have been arbitrary limited from -30° to +50° for the ankles, and from -90° to +150° for the hips. The white area represents the (static) stability region for upright stance, i.e., the region for which the center of mass is maintained above the feet. The height of the center of mass was manipulated by adding a 10 kg mass at the level of the knee or the neck, and the "length of the feet" was manipulated by changing the surface of support (See Appendix in Bardy et al. (1999) for details and equations).

A typical example is given on Figure 2, which illustrates for a set of participants (see Bardy et al., 1999 for details) the effect of changing the height of the center of mass (left Figure) or the length of the feet (right Figure) on (A) the stability region for stance and (B) the corresponding spontaneous values of the postural relative phase ϕ_{rel} for three amplitudes of the tracked target. As evidenced, a fixed value of a given constraint could produce different modes, depending on the value of other, independent constraints. For example, increasing the height of the center of mass yielded both in-phase and anti-phase modes, depending on the amplitude of target motion. An implication is that no single constraint is sufficient to predict the mode that will appear in a given situation. Previous analyses of multi-segment coordination in the control of stance have concentrated on physical properties of the observer-environment system. Such analyses have exemplified the influence of environmental properties (e.g., support surface length, motion, or compliance; Buchanan and Horak, 1999; Horak and Nashner, 1986), biomechanical properties of the body (foot length or body size; e.g., McCollum and Leen, 1989), or pathological deficits (e.g., Horak et al., 1990). The results reported on Figure 2 confirm that such factors can *shape* the coordination dynamics (e.g., Beek et al., 1995). However, they also demonstrate that these properties cannot fully determine or *specify* the coordination itself. Again, this is because a single mode was used for different biomechanical properties, and different modes were used for the same biomechanical property. It thus appears that postural coordination emerges out of the simultaneous (and sometimes competitive) influence of a variety of qualitatively different constraints.

4. Self-organization in the postural system

The emergence of only two postural modes of coordination, in-phase and anti-phase, out of the plethora of muscles and joints interacting together and creating a potentially very large number of multi-segmental coordination, is remarkable. It reveals that the high dimensionality of the postural system can be enslaved in, and captured by, a high order variable expressing the low-dimensional dynamics of that system: the relative phase between ankles and hips. Interestingly, the absolute value of ϕ_{rel} differed from 0° in these studies ($\phi_{rel} \approx 20\text{-}30°$), indicating that the ankles tended to lead the hips. This departure from pure in-phase mode-locking has not been observed in more or less identical components such as fingers or arms (e.g. Kelso, 1984; Schöner et al., 1986) but it has been observed in the context of whole-body coordination (Bardy et al., 1999; Buchanan and Horak, 1999). An explanation for this difference may be found in the frequency competition $\Delta\omega$ between the individual oscillators involved, as evidenced by classical work in inter-limb coordination (e.g., Schmidt et al., 1993; Sternad et al., 1992), demonstrating a systematic shift in the location of basins of attraction when $\Delta\omega$ differs from zero. The upper and lower parts of the body differing substantially in length, mass, and moment of inertia, non-similarities in their eigenfrequency may be responsible for the different value of relative phase, and for the variability of ϕ_{rel} during in-phase postural coordination.

More recently, we closely examined the self-organized features of these postural modes in the transient region, and discovered that the changes from in-phase to anti-phase (and from anti-phase to in-phase) exhibited characteristics of non-equilibrium phase transitions (Bardy et al., 2002). The same postural tracking task was used for this purpose. Standing participants were instructed to follow a moving target with their head. Target amplitude was kept constant (10 cm), but target frequency continuously increased or decreased between 0.05 Hz and 0.80 Hz in steps of 0.05 Hz, playing the role of the control parameter. The main results are summarized in Figures 3 and 4.

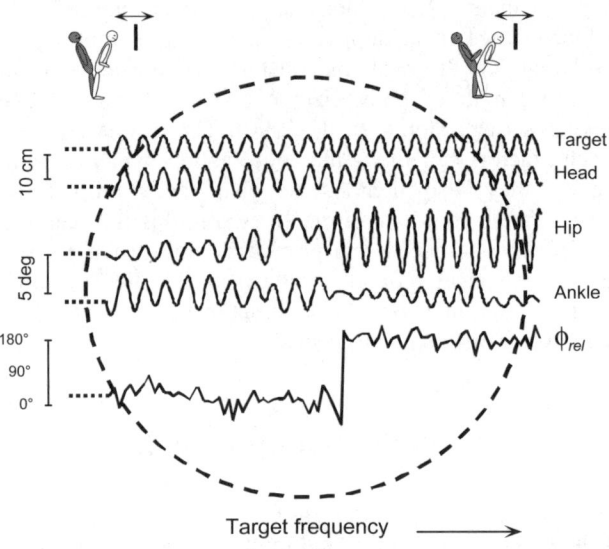

Figure 3. A zoom on the transition region for one typical record (*Up* condition) showing sustained in phase motion between the target and the head, and a transition from an in-phase to an anti-phase motion of the ankles and hips as target frequency is increased over time. This transition between patterns is highlighted in the change in relative phase ϕ_{rel} of the two joints. Adapted from Bardy et al. (2002).

As we increased (*Up* condition) or decreased (*Down* condition) the frequency at which the visual target moved, a frequency-induced loss of stability occurred, yielding *critical fluctuations* in the vicinity of the transition region. Transitions between in-phase and anti-phase modes were abrupt, with no intermediate state, and exhibited *hysteresis*: Transitions from in-phase to anti-phase indeed occurred at a higher frequency of target motion than transitions from anti-phase to in-phase. Finally, we applied an external perturbation, i.e., a shift in the direction in which the target was moving. These perturbations were applied either near to or far from the region in which transitions between modes were known to occur. Each mode was found to be less stable when perturbations were applied close to the transition region than when perturbations were applied far from it, as evidenced by a larger relaxation time values in the latter situation.

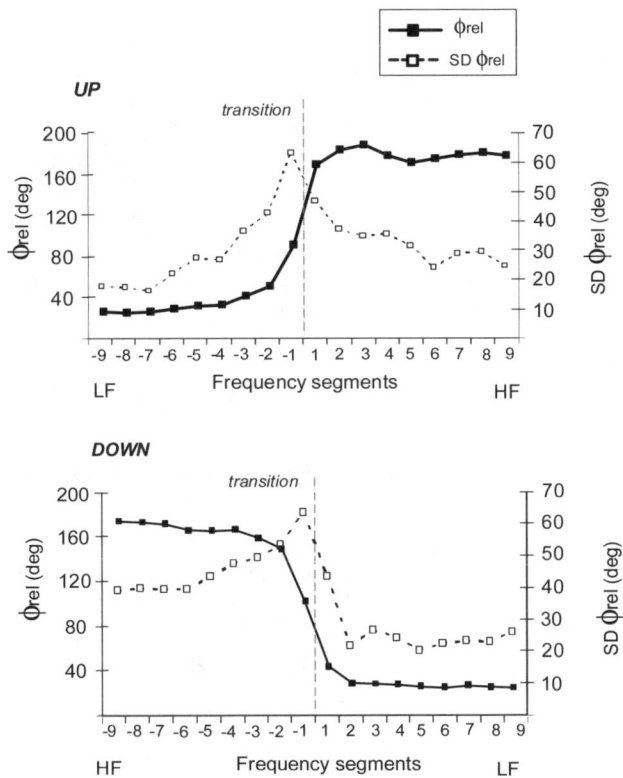

Figure 4. Postural transitions. Means and standard deviations of the point-estimate relative phase ϕ_{rel} (ten participants) as a function of target frequency in Up and Down conditions. Each frequency segment includes a temporal average of ϕ_{rel} over 4 cycles of oscillation, with an overlap of two cycles. *LF* and *HF* refer to low frequency and high frequency segments respectively. Adapted from Bardy et al. (2002).

These results have consequences for the development of a general theory of postural control. Consider for instance the well-developed neuromuscular approach according to which postural patterns are behavioral outcomes of centrally programmed neural strategies (e. g., Nashner and McCollum, 1985). Some of the results presented above are in concordance with this view. For example, the existence of two postural modes could be understood as resulting from two different neural plans for action. Similarly, abrupt transitions between those patterns could be explained in terms of a sudden shift between these plans. However, other aspects may be less easily explained, like the effects of hysteresis, critical fluctuations, and critical slowing down. Why would the central nervous system choose different postural modes for identical conditions? Why would the variability of relative phase increase when approaching the centrally-programmed changes in posture? What could be a realistic explanation for the existence of differential stability close and far from the transition region, as evidenced by the local relaxation time analysis? This type of questioning suggests that a general theory of postural transitions cannot be rooted in central mechanisms such as motor programs or solely in mechanical, energetic, or perceptual mechanisms constraining behavior (see Bardy et al., 2002). At a more general level, we think that any theoretical approach to posture could benefit from the present results, because the production and regulation of movement could take advantage of the non linearities outlined here (c.f., Rosenbaum, 1998). Indeed, these data provide evidences for the existence of self-organization in the postural system, and encourage further examination of the possibility that the interactions between the components of that system may be understood through the physics of non-equilibrium processes.

5. Supra-postural goals and coordination dynamics

A general characteristic of the studies reported here is that the postural coordination patterns (and the changes between them) are not prescribed by any type of instruction or intention, but emerge "for free" out of the interaction between the intrinsic dynamics of the postural system and a set of constraints acting upon it. This does not mean, however, that the goal of the task has no role to play in the emergence of postural patterns. Postural patterns are fundamentally influenced by the actor's intention (see Lee, 2003, for related arguments). Riccio and Stoffregen (1988) were among the first to suggest that postural stabilization is not an end in itself, but is valuable only to the extent that it facilitates the achievement of other goals. Stance can be controlled in different ways, which will differentially impact the performance of other behaviors. Thus, success of postural control actions may be most appropriately defined in terms of their impact on the achievement of supra-postural goals. It is meaningless to suggest that postural control is successful only if it minimizes postural sway. Minimal sway will facilitate the achievement of many goals, but not all possible goals. Some goals require body sway to be reduced, such as reading for instance, but other goals require body sway to be increased, such as following the target in our postural tracking task. In one exemplary study that tested the effect of task goal on body sway, Stoffregen et al. (2000) showed that standing participants instructed to search for letters in a text located at eye level exhibited less postural sway than when only instructed to look at a blank target placed at the same location. Among others, this result indicates that tasks that are super-ordinate to the control of posture, named *supra-postural tasks* (Stoffregen et al., 1999) are constraining the amount of sway in standing humans, and it is not unreasonable to assume that postural patterns may also be affected by the goal of the task.
In a different context, the literature on the perceptual control of balance has repeatedly demonstrated the importance of vision and visual coupling for the

maintenance of stance (e.g., Lee and Lishman, 1975; Schöner, 1991; Van Asten et al., 1988) but there have been almost no studies reporting the underlying postural coupling. The body is often considered as an inverted pendulum oscillating around the ankles and actively matching (Schöner, 1991), or passively driven by, the optical flow created by body sway. A classical result obtained in these studies is that coupling of postural sway to a moving visual surround decreases non linearly with increases in the frequency of imposed motion, or with increases in the distance to the visible surroundings (e.g., Dijkstra et al., 1992, 1994; Lestienne et al., 1977, Van Asten et al., 1988). One interpretation is that, because the natural frequency of the postural system can be different from the frequency of imposed motion, some decoupling effects between imposed and actual frequency can occur. To our knowledge there has been no analysis of the possible role of postural patterns (in-phase, anti-phase) to the observed changes in visual coupling. The differences that exist in sway amplitude and frequency for in-phase and anti-phase patterns suggests that changes in the amplitude and/or frequency of imposed motion might be accommodated by a change from one postural pattern to another. The shift between these modes could be expected to produce a temporary decoupling of the head from imposed optical flow.

By using the postural tracking paradigm, we recently addressed this issue. We tested the effect of the actor's intention on the emergence of postural patterns (Oullier et al., 2002). Participants standing in a moving room were instructed either to look at a target fixed on the front wall (*Looking* condition) or to follow the target with their head (*Tracking* condition). No instruction was given about what postural mode to adopt. Room frequency was increased or decreased in steps of 0.05 Hz between 0.10 Hz and 0.75 Hz. Dependent variables included the point-estimate relative phase between ankles and hips as well as gain and phase between motion of the head and motion of the target. Generally, the influence of task was greater than the influence of frequency, as evidenced by a decrease in both head's amplitude and head-target coupling in looking as compared to tracking. More interesting are the findings illustrated on Figure 5 that reveal several important features of the postural system.

First, a bi-modal distribution of relative phase values clearly appeared around 20° and 180°, indicating the same type of postural patterns for the two tasks, in-phase and anti-phase. Looking at or tracking a moving target are tasks that play a similar type of constrain on posture, which reveals that these two modes are stable attractors of the dynamics of the postural system. This indicates that the traditional inverted pendulum analogy may not be correct. Adaptive patterns of postural coordination underlie the simple act of looking at a target, as it does for tracking that target. Second, no intermediate state existed between in-phase and anti-phase, and a rapid jump existed for both tasks between in-phase and anti-phase, suggesting a similar two-attractors dynamics underlying looking and tracking. Third, Figure 5 also indicates a more broadly distributed range of relative phase values around 0° and 180° in looking as compared to tracking. This, together with the corresponding lower head-target coupling observed in Looking, argues for the existence of two sub-systems underlying the performance at the tracking task. A head-target system expressing the *visual coupling* with the environment, and a

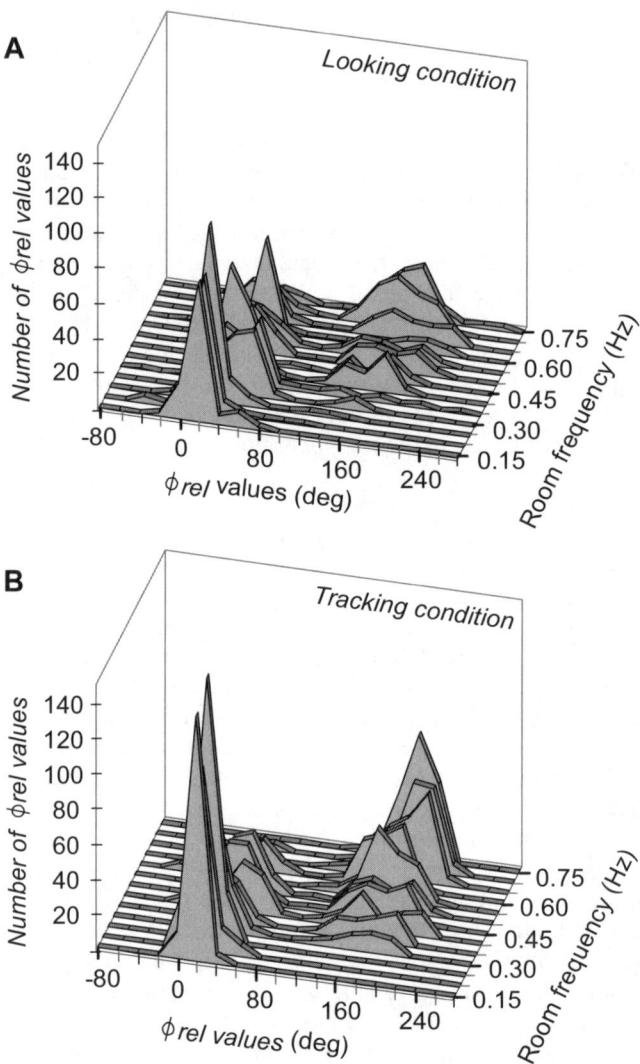

Figure 5. Effect of intention (A: looking at a target, B: tracking this target with the head) on the dynamics of postural patterns: Distribution for the two tasks of hip-ankle relative phase values as a function of room motion frequency. Adapted from Oullier et al. (2002).

hip-ankle sub-system expressing the *inertial coupling* between the two joints. The head-target system can be captured by the head-target relative phase (as well as cross-correlation and gain between and head and target). It is directly influenced by the goal of the task, which can be formalized in dynamical terms in the differential equation capturing that coupling (e.g., Schöner, 1989). The hip-ankle sub-system is captured by the hip-ankle relative phase. It is influencing, and being influenced by, the head-target coupling, in a circular causality. The possibility of modeling the dynamics of the postural system while taking into account the co-existence of these two interacting sub-systems, is now under investigation (Fourcade et al., 2002) an will be detailed shortly in the conclusion. For the present, we believe that these results offer converging evidence for the existence of self-organized phenomena operating at these two levels, and encourage examination of alternative interpretations of findings from studies of relations between posture and vision.

6. The dynamics of learning new postures

The process by which humans modify their spontaneous postural patterns to the requirement of new tasks is another important route to explore for the development of a general theory of postural patterns formation. In the context of stance-related behaviors in humans, various general accounts of changes between postural patterns due to learning have been advocated, mostly inspired by the work of Bernstein (1967). Bernstein's account of qualitative changes in movement organization emphasized the non-linear nature of the learning process. A common method for studying these non linearities is through the sequential reducing and releasing (also known as freezing and freeing, respectively) of the body's degrees of freedom (df). Learned movements exhibit changes in organization that correspond roughly to the three-stage model of motor learning proposed by Bernstein. Stage 1 consists of reducing the number of peripheral df to the minimum that can be controlled. In Stage 2, df are gradually released and incorporated in the coordination solution, while Stage 3 consists of exploiting the reactive phenomena existing at the interface between the organism and the environment (see Newell, 1996, for a review). Bernstein's theory, although not yet fully tested (Newell and Vaillancourt, 2001), has proven to be a robust conceptual framework for describing behavioral changes observed during the acquisition of a wide variety of motor skills involving the postural system, such as dart throwing (McDonald et al., 1989), pistol shooting (Arutyunyan et al., 1968), or ski-like body oscillating (Vereijken et al., 1992).

In addition, experimental and theoretical studies have provided evidence for the operation of self-organizational principles in the domain of learning new human inter-limb coordination (e.g., Fontaine et al., 1997; Zanone and Kelso, 1992, 1997; Verschueren et al., 1997; Wenderoth and Bock, 2001). At a theoretical level, Schöner (1989, Schöner et al., 1992) has postulated that changes due to learning emerge out of the cooperative or competitive interplay between the initial, well-learned pattern of coordination (the so-called *intrinsic dynamics*) and the dynamic characteristics of the to-be-learned pattern, or *behavioral information*.

Experimental results were consistent with this proposal (Zanone and Kelso, 1992, 1997). It is to be noted, however, that most of these studies have concentrated on coordination of individual fingers, joints, or arms, and that empirical data in the field of posture or stance are rare. One reason may be that it is not easy to measure how a complex multi-joint system like a human body changes over time. Another reason is that there may be differences in the time scales at which long-term changes occur in the bimanual vs. postural patterns. Whereas the acquisition of a new bi-manual pattern sometimes can take less than fifty trials, learning and stabilizing new postures often involves years of practice, as illustrated by many examples in sport. Preliminary information about changes in postural coordination due to learning has been obtained, however, in a recent study using the postural tracking task (Marin, Bardy and Bootsma, 1999). Increasing the frequency of target oscillation produced a transition from the in-phase mode to the anti-phase mode. However, this transition occurred earlier for non-gymnasts than for gymnasts, suggesting that the non-gymnasts might be able to learn to maintain the in-phase mode at higher frequencies. The delay in phase transition in gymnasts was attributed to an increase in both their global muscular stiffness, resulting from the co-activation of posterior and anterior muscles acting at the main joints, and in the gain of the stretch reflex (Fitzpatrick et al., 1992). In addition, expertise was found to interact with other constraints (e. g., those defined by the task and the properties of the surface of support), yielding specific coordination modes adapted to the rules of the tracking task.

Based on these earlier results, we recently conducted a series of experiments in which the design and analysis allowed to assess the changes due to learning as the result of the competitive interplay between the initial pattern of coordination and the characteristics of the to-be-learned pattern (Bardy, Faugloire et al., 2003; Faugloire & Stoffregen, 2003). In these experiments, standing participants were instructed to learn a new phase relation between the ankles and the hips. The body reorganization during the acquisition of this new postural pattern, as well as the destabilization of the intrinsic dynamics consecutive to learning, was assessed. The methodology involved a pre-test/learning/post-test design. The pre-test and the post-test involved the postural tracking task, with target motion of constant amplitude (10 cm) and various frequencies (from 0.25 Hz to 0.65 Hz). In the learning period, participants were instructed to learn a new phase relation between ankles and hips, namely 135 degrees. They did not execute the tracking task. They stood in the middle of the laboratory, looking straight ahead with the arms crossed in the back, and moved their body in a way to reproduce the requested relative phase pattern, at a natural, i.e., not constrained, amplitude and frequency. 30 trials consisting of 10 body oscillations were recorded for each participant. A feedback was given to them every 3 trials, consisting of a state space (an hip-ankle plane) illustrating the discrepancy between the performed and the desired (135°) postural pattern. Figure 6 illustrates the main results for a group of 14 participants (Bardy et al., 2003). Although there were differences between participants in terms of learning and rate of destabilization, learning a postural relative phase of 135 degrees did occur, as evidenced by the migration over trials of the current relative phase in the direction of the to-be learned pattern, by the decrease in its standard deviation (Figure 6, middle), and by the decrease in movement time. Moreover, an interesting finding was the destabilization effect due to learning, witnessed by the difference in both the spontaneous relative phase (here 180 degrees) and its

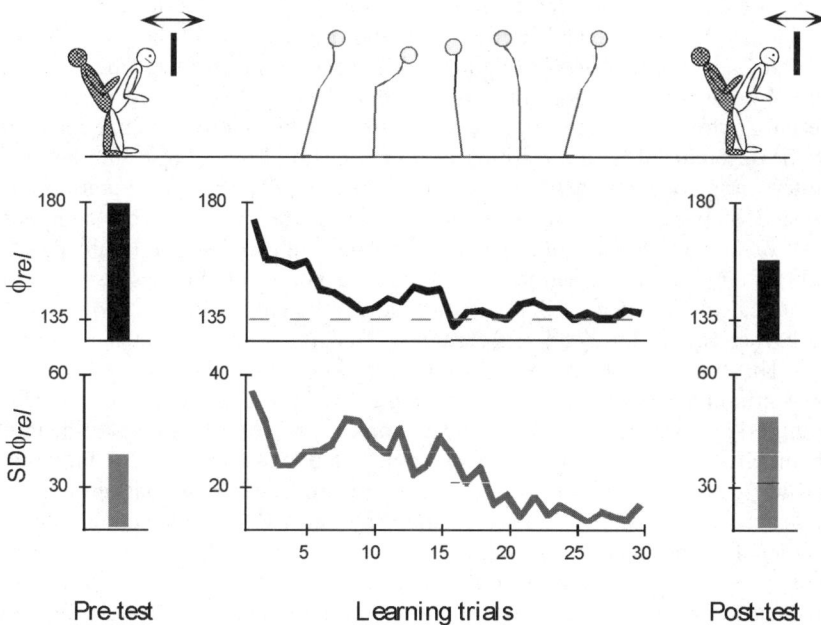

Figure 6. The effects of learning a new relative phase (135°) on the stability of a spontaneous postural pattern (180°). Learning is evidenced by a change in the relative phase φrel toward 135° and a decrease in its variability (middle). The destabilization due to learning is evidenced by the change in initial φrel in the direction of the to-be-learned pattern between the pre-test (left) and the post-test (right), as well as in the increase in its variability SDφrel. Adapted from Bardy et al. (2003).

variability between the pre-test and the post-test. Overall, participants exhibited a postural coordination at the post-test that was in between the initial, spontaneous, pattern (180 degrees) and the learned pattern (135 degrees). Complementary findings were recently obtained in the examination of the links between sport expertise and the learning of a novel pattern postural coordination, reflecting the competition between the dynamics of the novel and pre-existing patterns (Ehrlacher et al., 2003). Although preliminary, these results strongly suggest that learning a new, non-spontaneous, postural pattern requires the reconfiguration of the intrinsic dynamics of the postural control structure compatible with the operating constraints, and the discovery (and stabilization) of a new equilibrium between the constituents of the movement performed.

7. Conclusion

We believe that the postural tracking task provides a general and useful paradigm for investigating the non linear dynamics of human posture. The main signatures of self-organization observed with this paradigm (including critical fluctuations, bifurcation, hysteresis, and critical slowing down) should be considered as important features that need to be taken into account in a general theorization of postural control. In addition, the findings that these self-organized modes are (i) emerging out of a coalition of intrinsic, environmental, and intentional constraints, and (ii) destabilized by the learning of a new postural mode, open the possibility of integrating concepts such as perception, learning, expertise, or intention, in the language of postural dynamics. Recently, we created a mixed, or 'compound', model of human posture that captures these non-linearities (see Fourcade et al., 2003, for a preliminary treatment). The model is composite in the sense that it is a mechanical model, that is linking joints and segments, with units of mass and length, that can produce the self-organized signatures observed at the behavioral level. The human body is represented with two non-deformable segments, one representing the head-trunk system rotating around the hips, the other representing the thigh-legs system rotating around the ankles, in a double-inverted-pendulum system. Stiffness and damping terms acting at the two joints have biologically plausible values. Oscillations of the body are produced by a constant-amplitude torque acting at the ankles, counterbalanced by a resistive torque acting at the hips in order for the feet to be kept in contact with the ground. Interestingly, simulations of the double pendulum indicate the presence of two hip-ankle coordinative patterns accompanying the continuous increase of target frequency (in phase, anti-phase), as well as the presence, with appropriate initial conditions, of critical fluctuations, bifurcation, hysteresis, and critical slowing down. Thus, this accuracy suggests that it is possible to root the general organizational principles accompanying movement control into the biomechanical (or neurophysiological) substrates of specific biological systems, such as the postural system. It also suggests that an intermediate position between pure structural models —searching for structural and causal explanations of behaviors — and pure phenomenological models —searching for abstract and general principles of (self-)organization (c.f., Beek et al., 1995) can be useful for modeling human posture and movement.

Acknowledgments

The research reported in this article was supported by grants from the *Centre National de la Recherche Scientifique* (CNRS/NSF-3899), with additional support from the University of Paris XI and the *Institut Universitaire de France*. I would like to thank my former and current students Ludovic Marin, Olivier Oullier, Elise Faugloire, Caroline Ehrlacher, Cédric Bonnet, as well as Tom Stoffregen, Reinoud Bootsma, and Paul Fourcade for their contribution to portions of this work.

Correspondence concerning this article should be addressed to Benoît G. Bardy, Research Center in Sport Sciences, University of Paris Sud XI, Bâtiment 335, 91405 Orsay Cedex, France. Electronic mail may be sent via the Internet to benoit.bardy@staps.u-psud.fr.

References

Allum JH, Honegger F, Schicks H (1993) Vestibular and proprioceptive modulation of postural synergies in normal subjects. J Vestib Res 3, 59-85

Arutyunyan RH, Gurfinkel VS, Pirskii ML (1968) Investigation of aiming at a target. Biophysics 13, 536-538

Bardy BG, Marin L (1997) Pour une approche fonctionnelle des coordinations posturales [For a functional approach to postural coordination]. In: Lacour M (ed.) Posture et equilibre: Pathologies, vieillissement, stratégies, modélisation. Sauramps Medical, Montpellier, 139-154

Bardy BG, Faugloire E, Stoffregen, TA (2003) The dynamics of learning new postural patterns. Article submitted

Bardy BG, Marin L, Stoffregen TA, Bootsma RJ (1999). Postural coordination modes considered as emergent phenomena. J Exp Psychol Human 25, 1284-1301

Bardy BG, Oullier O, Bootsma RJ, Stoffregen, TA (2002) The dynamics of human postural transitions. J Exp Psychol Human 28, 499-514

Beek PJ, Peper CE, Stegeman DF (1995) Dynamical models of movement coordination. Hum Movement Sci 14, 573-608

Bernstein N (1967) The co-ordination and regulation of movement. Pergamon Press, Elmsford, NY

Buchanan JJ, Horak FB (1999) Emergence of postural patterns as a function of vision and translation frequency. J Neurophysiol 81, 2325-2339

Buchanan JJ, Kelso JAS, DeGuzman GC (1997) Self-organization of trajectory formation, I. Experimental evidence. Biol Cybern 76, 257-273

Carson RG, Goodman D, Kelso JAS, Elliott D (1995) Phase transitions and critical fluctuations in rhythmic coordination of ipsilateral hand and foot. J Motor Behav 27, 211-224

Corna S, Tarantola J, Nardone A, Giordano A, Schieppati M (1999) Standing on a continuously moving platform: is body intertia counteracted or exploited? Exp Brain Res 124, 331-341

Diedrich FJ, Warren WH (1995) Why change gaits? Dynamics of the walk-run transition. J Exp Psychol Human 21, 183-202

Dijkstra TMH, Gielen, CCAM, Melis BJM (1992) Postural responses to stationary and moving scenes as a function of distance to the scene. Hum Movement Sci 11, 195-203

Dijkstra TMH, Schöner G, Gielen, CCAM (1994) Temporal stability of the action-perception cycle for postural control in a moving visual environment. Exp Brain Res 97, 477-486

Ehrlacher C, Bardy BG, Faugloire E, Stoffregen TA (2003) Sports expertise influences learning of postural coordination. In: Rogers S, Effken J (eds.) Studies in perception and action VII. Erlbaum, Hillsdale, in press

Faugloire E, Stoffregen TA (2003) The dynamics of learning a new posture. In: Rogers S, Effken J (eds.) Studies in perception and action VII. Erlbaum, Hillsdale, in press

Fitzpatrick RC, Taylor JL, McCloskey DI (1992) Ankle stiffness of standing humans in response to imperceptible perturbation, reflex and task-dependent components. J Physiol 454, 533-547

Fontaine RJ, Lee TD, Swinnen SP (1997) Learning a new bimanual coordination pattern, reciprocal influences of intrinsic and to-be-learned patterns. Can J Exp Psychol 51, 1-9

Fourcade P, Bardy BG, Bonnet C (2003) Modelling human postural transitions. In: Rogers S, Effken J (eds.) Studies in perception and action VII. Erlbaum, Hillsdale, in press

Haken H, Kelso JAS, Bunz H (1985) A theoretical model of phase transitions in human hand movements. Biol Cybern 51, 347-356

Horak FB, Nashner LM (1986) Central programming of postural movements, adaptation to altered support-surface configuration. J Neurophysiol 55, 1369-1381

Horak FB, Nashner LM, Diener H C (1990) Postural strategies associated with somatosensory and vestibular loss. Exp Brain Res 82, 167-177

Jeka JJ, Schöner G, Dijkstra TMH, Ribeiro P, Lackner JR (1997) Coupling of fingertip somatosensory information to head and body sway. Exp Brain Res 113, 475-483

Kelso JAS (1984) Phase transitions and critical behavior in human bimanual coordination. Am J Physiol-Reg I 15, 1000-1005

Kelso JAS, Buchanan JJ, Wallace SA (1991) Order parameters for the neural organization of single, multijoint limb movement patterns. Exp Brain Res 85, 432-444

Lee TD (2003) Intention in bimanual coordination performance and learning. This volume

Lee DN, Lishman R (1975) Visual proprioceptive control of stance. J Hum Movement Stud 1, 87-95

Lestienne F, Soechting J, Berthoz A (1977) Postural readjustments induced by linear motion of visual scenes. Exp Brain Res 28, 363-384

Marin L, Bardy BG, Bootsma RJ (1999) Level of gymnastic skill as an intrinsic constraint on postural coordination. J Sport Sci 17, 615-626

Marin L, Bardy BG, Baumberger B, Flückiger M, Stoffregen TA (1999) Interaction between task demands and surface properties in the control of goal-oriented stance. Hum Movement Sci 18, 31-47

McCollum G, Leen TK (1989) Form and exploration of mechanical stability limits in erect stance. J Motor Behav 21, 225-244

McDonald PV, van Emmerik REA, Newell KM (1989) The effects of practice on limb kinematics in a throwing task. J Motor Behav 21, 245-264

Nashner LM, McCollum G (1985) The organization of postural movements, A formal basis and experimental synthesis. Behav Brain Sci 26, 135-172

Nashner LM, Shupert CL, Horak FB, Black FO (1989) Organization of posture control, A analysis of sensory and mechanical constraints. Prog Brain Res 80, 411-418

Newell KM (1985) Coordination, control and skill. In: Goodman D, Wilberg RB, Franks IM (eds.) Differing perspectives in motor learning, memory, and control. North-Holland, Amsterdam, 295-317

Newell, KM (1996) Change in movement and skill, learning, retention, and transfer. In: Latash M, Turvey MT (eds.) Dexterity and its development. Erlbaum, Hillsdale, 63-84

Newell KM, Vaillancourt DE (2001) Dimensional change in motor learning. Hum Movement Sci 20 695-715

Oullier O, Bardy BG, Stoffregen TA, Bootsma RJ (2002) Postural coordination in looking and tracking tasks. Hum Movement Sci 21, 147-167

Paï Y-C, Patton J (1997) Center of mass velocity-position predictions for balance control. J Biomech 30, 347-354

Riccio GE, Stoffregen TA (1988) Affordances as constraints on the control of stance. Hum Movement Sci 7, 265-300

Rosenbaum DA (1998) Is dynamical systems modeling just curve fitting? Motor Control 2, 101-104

Saltzman EL, Kelso JAS (1985) Synergies, stabilities, instabilities and modes. Behav Brain Sci 8, 161-163

Schmidt RC, Carello C, Turvey MT (1990) Phase transitions and critical fluctuations in the visual coordination of rhythmic movements between people. J Exp Psychol Human 16, 227-247

Schmidt RC, Shaw BK, Turvey MT (1993) Coupling dynamics in interlimb coordination. J Exp Psychol Human 19, 397-415

Schöner G (1989) Learning and recall in a dynamic theory of coordination patterns. Biol Cybern 62, 39-54

Schöner G (1991) Dynamic theory of action-perception patterns, The "moving room" paradigm. Biol Cybern 64, 455-462

Schöner G, Haken H, Kelso JAS (1986) A stochastic theory of phase transitions in human hand movement. Biol Cybern 53, 247-257

Schöner G, Zanone PG, Kelso JAS (1992) Learning as change of coordination dynamics. Theory and experiment. J Motor Behav 24, 29-48

Sternad D, Turvey MT, Schmidt RC (1992) Average phase difference theory and 1:1 phase entrainment in interlimb coordination. Biol Cybern 67, 223-231

Stoffregen TA, Pagulayan RJ, Bardy BG, Hettinger LJ (2000) Modulating postural control to facilitate visual performance. Hum Movement Sci 19, 209-220

Stoffregen TA, Riccio GE (1988) An ecological theory of orientation and the vestibular system. Psychol Rev 95, 3-14

Stoffregen TA, Smart LJ, Bardy BG, Pagulayan RJ (1999) Postural stabilization of looking. J Exp Psychol Human 25, 1641-1658

Van Asten WNJC, Gielen CCAM, Denier van der Gon JJ (1988) Postural adjustments induced by simulated motion of differently structured environments. Exp Brain Res 73, 371-383

Vereijken B, van Emmerik REA, Whiting HTA, Newell KM (1992). Free(z)ing degrees of freedom in skill acquisition. J Motor Behav 24, 133-142

Verschueren SM, Swinnen SP, Dom R, De Weerdt W (1997) Interlimb coordination in patients with Parkinson's disease, motor learning deficits and the importance of augmented information feedback. Exp Brain Res 113, 497-508

Wenderoth N, Bock O (2001) Learning of a new bimanual coordination pattern is governed by three distinct processes. Motor Control 5, 23-35

Woollacott MH, Jensen JL (1996) Posture and locomotion. In: Heuer H, Keele S (eds) Handbook of perception and action; Vol. 2, Motor Skills. Academic Press, London, 333-403

Yoneda S, Tokumasu K (1986) Frequency analysis of body sway in the upright posture. Acta Oto-laryngol 102, 87-92

Zanone PG, Kelso JAS (1992) Evolution of behavioral attractors with learning, Nonequilibrium phase transitions. J Exp Psychol Human 18, 403-421

Zanone PG, Kelso JAS (1997) Coordination dynamics of learning and transfer, Collective and component levels. J Exp Psychol Human 23, 1454-1480

Noise Associated with the Process of Fusing Multisensory Information

John Jeka[1,2] & Tim Kiemel[3]

1 -Program in Neuroscience & Cognitive Science, 2 - Department of Kinesiology,
3 -Department of Biology, University of Maryland, College Park, MD

1. Introduction

The small, continuous displacements around the actual vertical upright that characterize upright stance behavior reflect a complex control process that involves estimation of body position heavily dependent upon the integration of information from multiple sensory systems (for reviews, see Dietz, 1992; Horak & Macpherson, 1996; Nashner, 1981). Precise estimation of self-orientation is particularly important when confronted with new environmental conditions which require an immediate updating of the relative importance of different sources of sensory information (i.e., sensory reweighting). For example, environmental changes such as moving from a light to a dark environment or from a fixed to a moving support surface (e.g., onto a moving walkway at the airport) require an updating of sensory weights to current conditions so that muscular commands are based on the most precise and reliable sensory information available (Teasdale et al., 1991; Wolfson et al., 1985; Woollacott et al., 1986). On a longer time scale, sensory reweighting is also crucial when an individual sensory input is lost due to injury, disease or aging (e.g., proprioceptive loss due to diabetic peripheral neuropathy, vestibular dysfunction from Meniere's disease, etc.).

Despite a long list of proposed models in the literature, it is arguable that models have had little impact on our understanding of postural control, particularly in terms of multisensory integration. For example, existing models that represent the role of the three primary sensory systems for balance (visual, vestibular and somatosensory) can reproduce the experimental observation that adding sensory information reduces the total sway variance of quiet stance (e.g., Schöner, 1991; van der Kooij et al., 1999). However, since virtually any proposed mechanism for using sensory information would have this property, it does not help us identify the mechanism actually used by the postural control system. Hence, reduction in mean sway amplitude due to additional sensory information is not a particularly useful property. Additional properties/constraints are necessary if modeling is to be used to understand the mechanisms underlying the estimation and control of posture. Here we illustrate how new properties can be identified with well-known but under-utilized (at least in the posture literature) descriptive techniques which then serve to constrain the form of a mechanistic model of postural control.

2. Developing a mechanistic model

Mechanistic Models. Mechanistic models explicitly refer to underlying mechanisms that are known or hypothesized to contribute to postural control. The most straightforward approach in developing a mechanistic model of postural control is to characterize what is known about underlying physiological subsystems (e.g., transfer functions of sensory subsystems, see Borah, 1988) and then to use theoretical concepts from dynamical systems theory (Dijkstra et al., 1994; Giese et al., 1996; Jeka et al., 1997; 1998; Schöner, 1991) or control theory (e.g., Johansson et al, 1988; Kuo, 1995; van der Kooij et al, 1999; 2001; Peterka, 2000) to "glue" the individual subsystems together. These models have been able reproduce certain characteristics of experimental postural sway data, although to a limited degree. For example, van der Kooij et al. (2001) recently presented a mechanistic model with a sophisticated scheme for sensory reweighting. Even though they were able to account for data on the postural control of individuals with bilateral vestibular loss (Perterka & Benolken,1995), there were inconsistencies. For instance, the model predicted that bilateral vestibular loss patients will fall when standing on a sway-referenced surface with eyes open in a stationary visual environment. Many studies have shown the contrary (e.g., Black et al., 1982; Nashner et al., 1982). In short, there is no mechanistic model of postural control that can account for the postural sway behavior of healthy subjects as well as a patient population such as individuals with vestibular loss. A mechanistic model is ultimately the goal because of the link to underlying physiological subsystems. However, we argue that before any mechanistic model is considered feasible it must be able to account for the dynamic characteristics identified by a descriptive model.

3. Descriptive approaches

Ever since Begbie (1967) first suggested that postural sway consists of two fundamental components, a slow oscillatory component between 0.5-1 Hz and one a fast component between 1.5-2.5 Hz, different schemes have been proposed to decompose the postural sway trajectory. The idea is that knowledge of such fundamental components may provide insight towards the structure of the control system underlying standing posture, which remains poorly understood.

4. Model free

There are two approaches to describe the dynamics of sway trajectories: model-free or model-based. The most common model-free descriptors are the power-spectral-density function, the autocovariance function and the diffusion function (Collins & DeLuca, 1993). These three functions contain the same information if the output process is stationary and completely characterizes the process if it is also linear. For example, Collins & Deluca (1993) implemented stabilogram diffusion analysis to characterize center of pressure trajectories as a correlated random walk. Their analysis led them to hypothesize a short-term and long-term regime in COP trajectories (for review Riley et al., 1999). More recently, Zatsiorsky and Duarte (1999) identified slow and fast components in COP trajectories through an instant equilibrium point (IEP) analysis which separates center of pressure trajectories into rambling and trembling components.

Typically, these descriptive accounts interpret their findings mechanistically, by linking specific components to underlying physiological mechanisms. For example, Begbie (1967) speculated that the fast oscillations were due to stretch reflex pathway characteristics which enhanced somatosensory function when visual or vestibular were disrupted. Slow oscillations emerged from sway corrections based upon vestibular information. Collins and Deluca (1993) attributed the slow and fast regions of the stabilogram diffusion plots to different regimes of nervous system control. The short-term region corresponded to an open-loop form of control and the long-term region corresponded to a closed-loop control form of control. Zatsiorsky and Duarte (1999) made no mechanistic claims from their IEP analysis. One possibility is a model developed by Dijkstra (2000). He proposed a 3rd order dynamic set point model which coupled the slowly moving equilibrium position to the faster oscillations around that dynamic equilibrium that could be considered similar to the rambling component. The question remains whether the set point is explicitly controlled by the nervous system, albeit somewhat imprecisely, or whether the variability in the set point reflects a more passive form of control due to the inherent mechanical properties of the musculoskeletal system.

Model-free descriptions have an advantage in that they are based directly on experimental trajectories and are straightforward to compute. However, a problem is that it is not clear how the model-free decompositions relate to any given descriptive or mechanistic model of postural sway. For example, numerous models have been proposed to account for the stabilogram diffusion findings. Collins & Deluca (1993) originally proposed a 5th order model, consisting of two linearly superimposed random walkers bounded by threshold-based alternating springs. Newell et al (1997) proposed an alternative 1st order model which accounted for the majority of center of pressure variance, calling into question whether two distinct control regimes existed. Peterka (2000) then showed that a PID control model could reproduce an entire family of stabilogram diffusion plots, many of which were similar to those illustrated by Collins and Deluca (1993). Because the PID model was linear, it questioned the existence of two distinct regions of control. Another scheme proposed by Chow & Collins (1995) suggested that the stabilogram-diffusion findings could be modeled as an elastic pinned-polymer under the influence of noise.

5. Model based

Descriptive models are those whose form and parameters are determined from fitting experimental trajectories (i.e., time series). The goal of descriptive modeling is to find a model with the minimal number of parameters that characterizes the observed behavior to a reasonable degree of accuracy, in our case, postural sway. Since postural sway trajectories measured under the same experimental conditions vary from trial to trial, sway trajectories are viewed as the output of a stochastic process, that is, a process whose output depends on random (as well as deterministic) factors. Therefore, descriptive models of postural sway are stochastic models whose output, as with postural sway itself, varies from trial to trial.

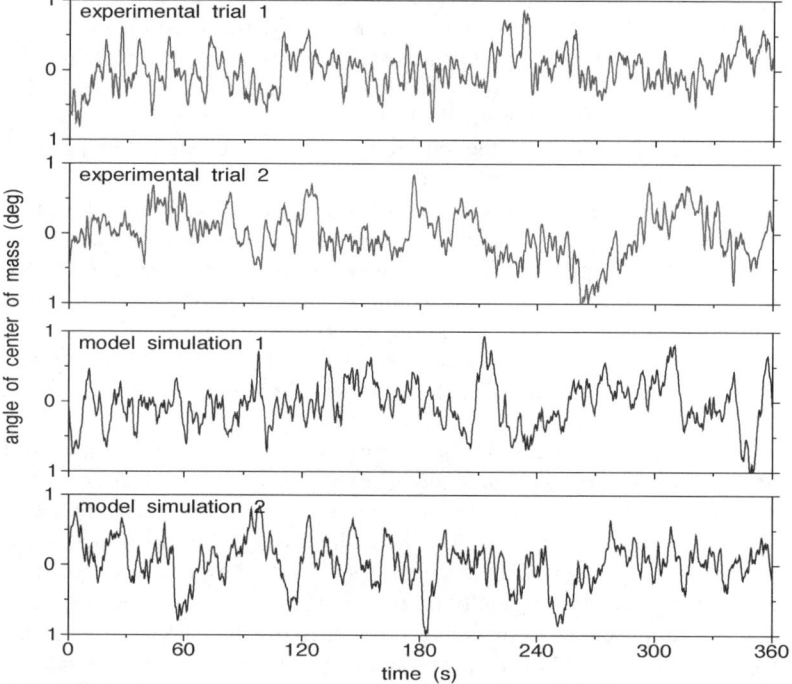

Figure 1. Experimental postural sway trajectories and simulated trajectories produced by our six-parameter time series (ARMA) model.

Comparisons between a model and experimental postural sway are based on statistical properties, such as the autocovariance function, not on the exact shape of trajectories from individual trials. Figure 1 shows two trajectories produced from actual postural sway and two trajectories from a descriptive model we have developed (Kiemel et al., 2002). A look at the four trajectories illustrates that it would be very difficult to judge which is actual sway and which is produced from a model. Even though they are not exact copies, our analysis has shown that they share the same basic statistical features (Kiemel et al, 2002).

The most common model-based technique is to select an auto-regressive moving-average (ARMA) model that approximates the postural dynamics (Johannson et al., 1988; Kiemel et al., 2002; Oie et al., 2002; Fransson et al., 1998). An ARMA model can be used to completely characterize the output of any finite-dimensional linear process. An ARMA model is an example of a descriptive model, because it does not explicitly refer to any mechanism in the underlying stochastic process. Its form and parameters are determined based on experimental trajectories.

One advantage of using a descriptive model to characterize postural sway is that the model has only a few parameters. Therefore, to study how postural sway varies with experimental condition, it suffices to study how these parameters vary with condition. Another advantage of a descriptive model is that it clearly and concisely identifies the dynamic characteristics of postural sway that must be accounted for by any "mechanistic" model. For example, the structure of the ARMA model provides a minimum on the number of variables necessary to model the underlying system. In the case of Figure 1, the structure of the ARMA model implies that at least three variables are necessary to model the postural control system. Moreover, the ARMA model gives information about the rate constants of the underlying system's dynamics in the form of eigenvalues. A third advantage is that because the descriptive model is based on fitting actual sway trajectories, it accounts for all contributions to the postural system. For instance, if a descriptive model reproduces the statistical properties of the experimental trajectories, then it is assured that a vestibular contribution to the overall variance is *implicitly* included in the model, even though there is no term in the model that represents vestibular information. In order to represent the vestibular contribution *explicitly*, the step to a mechanistic model is required.

Below we show how descriptive techniques aid in distinguishing different models of postural control as well as identify new mechanisms that are important. A detailed mathematical description of these techniques can be found in Kiemel et al (2002).

6. Linking descriptive and mechanistic models

One of the most common findings in the postural literature is that additional sensory information leads to a reduction in the mean amplitude of postural sway. For example, we and others have shown that additional light touch or visual information reduces postural sway amplitude (Clapp & Wing, 1999; Dickstein et al., 2001; Holden et al., 1984; Jeka & Lackner, 1994; 1995; Reginella et al., 1999; Riley et al., 1997). Unfortunately, this is not a useful empirical finding to distinguish different types of models that describe the use of sensory information to control posture because most models predict that adding sensory information will decrease sway variance (e.g., Schöner, 1991; van der Kooij et al., 1999). Therefore, empirical characteristics other than mean sway amplitude are necessary to test the relative merit of different theoretical frameworks. To identify such characteristics, we fitted time series models to experimental sway trajectories and then used these characteristics to test and modify existing models of postural control (Kiemel et al., 2002).

We first ran a simple experiment in which nine subjects stood quietly while available sensory information was manipulated. Figure 2 shows the experimental setup. Subjects stood in a tandem stance in front of a computer generated visual display while maintaining right index fingertip contact with a stationary plate that measured the applied forces. Ultrasound receivers measured head and approximate center of mass kinematics. An auditory alarm sounded if above threshold forces were applied, signaling to the subject to reduce applied force without losing contact of the plate. In general, the task is easy for subjects. After one practice trial, subjects rarely set off the alarm. The visual display consisted of 140 0.2 x 0.2 deg stationary dots projected randomly across a translucent 2.5 m x 2 m screen. Subjects wore goggles which limited their field of view to just the field of random dots. The four sensory conditions were: i) no touch - eyes closed, in which subjects stood with arms hanging loosely by their side; ii) light touch – eyes closed ; iii) no touch – eyes open; and iv) light touch –eyes open. Complete experimental details can be found in (Kiemel et al., 2002).

For each of the nine subjects and for each of the four sensory conditions, we used the subject's postural sway trajectories to fit parameters in linear stochastic models of orders 1 to 8. We then selected the appropriate order based on likelihood-ratio tests. The goal was to find a model whose postural sway trajectories were statistically similar to the experimental trajectories. Because a stationary linear process is completely characterized by its autocovariance function, rather than reporting the parameters of an ARMA model, one can simply report the parameters of its autocovariance function. The autocovariance function describes the average relationship between a point at time t with a point at some future time $t + \tau$. In most cases, all but three terms in autocovariance function were small, and the autocovariance function could be approximated by

$$E[X(t)X(t+\tau)] = \kappa_r e^{\lambda_r |\tau|} + \kappa_c e^{\lambda_c |\tau|} + \overline{\kappa}_c e^{\overline{\lambda}_c |\tau|} \quad (1)$$

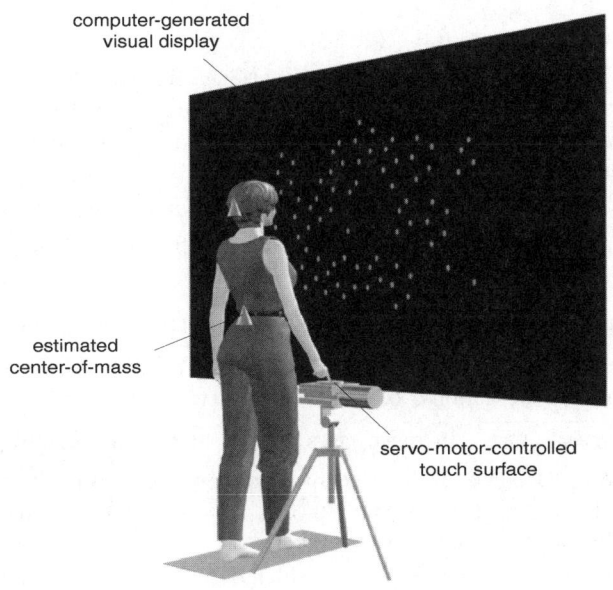

Figure 2. Subject standing in tandem stance in the multi-sensory vision-touch apparatus.

The parameters of the autocovariance function are the eigenvalues, λ_r, λ_c and $\bar{\lambda}_c$ and their corresponding coefficients κ_r, κ_c and $\bar{\kappa}_c$. All statistical properties of the model's sway trajectories can be expressed in terms of these six parameters. For example, the variance of the trajectories is $\kappa_{tot} = \kappa_r + \kappa_c + \bar{\kappa}_c$. Eigen values tell you the rate at which the autocovariance function decays with increasing time delay. One eigenvalue λ_r describes a slow 1st-order decay component of the autocovariance function. The other two eigenvalues λ_c and $\bar{\lambda}_c$ account for a faster-decaying damped-oscillatory component.

Note that we speak here of decomposing the autocovariance function of the model, not its sway trajectories. It is also possible to decompose postural sway trajectories into slow-decay and damped-oscillatory components. However, such decomposition is not unique.

The decomposition of the autocovariance function into a first-order decay component and a damped-oscillatory component allows definition of the following six measures:

- The slow-decay rate $\beta = -\lambda_r$, which describes how quickly the first-order decay component of the autocovariance function decays with time delay τ. Note that based on its definition, the slow-decay rate β is not necessarily slow. The term "slow-decay rate" is based on the experimental results from Kiemel et al. (2002).
- The damping $\alpha = -(\lambda_c + \bar{\lambda}_c)$, which describes how quickly the damped-oscillatory component of the autocovariance function decays with time delay τ. The rate constant of the decay is $\alpha/2$.
- The eigenfrequency $\omega_0 = -\sqrt{\lambda_c \bar{\lambda}_c}$, which is the approximate angular frequency of the damped-oscillatory component if α is small.
- The standard deviation σ of the model's sway trajectories. Typically, $\sigma \approx \sqrt{\kappa_r + \kappa_c + \bar{\kappa}_c}$. In most cases, σ^2 is also approximately equal to the average variance of the three sway trajectories used to fit the ARMA model. However, slow trends in the data that are not modeled as stochastic variation do not contribute to σ.
- The slow-decay fraction κ_r / κ_{tot}, which describes the relative size of the slow-decay component of the autocovariance function.
- The damped-oscillatory fraction $2|\kappa_c|/\kappa_{tot}$, which describes the relative size of the damped-oscillatory component of the autocovariance function.

Because the absolute value of κ_c is used in the definition of the damped-oscillatory fraction, the sum of the slow-decay and damped-oscillatory fractions can be greater than 1. Therefore, these measures cannot be simply interpreted as a partition of the sway variance. However, roughly speaking, if the slow-decay fraction is near 1 and the damped-oscillatory fraction is near 0, then the slow-decay component of postural sway accounts for most of the sway variance.

Figure 3 illustrates how these measures behaved across sensory condition. Figures 3a-c show measures characterizing the eigenvalues; Figures 3d-f show measures characterizing the coefficients. Figure 3d shows the typical result of increased standard deviation of sway amplitude as sensory information is removed. Despite this change in sway standard deviation across condition, none of the other measures showed a significant dependence on sensory condition. This means that adding vision or touch information did not result in any significant change in the overall stochastic structure of the control system. Sway standard deviation decreased because additional sensory information allows for a more precise estimate of center of mass dynamics and consequently, more precise control.

Figure 3e shows that the slow-decay fraction remained near 1 for all the sensory conditions. Figure 3f shows that the damped-oscillatory fraction remained relatively low across all the sensory conditions. Thus, roughly speaking, the slow-decay process accounted for most of the sway variance in each sensory condition, even though the absolute levels of sway differed across conditions.

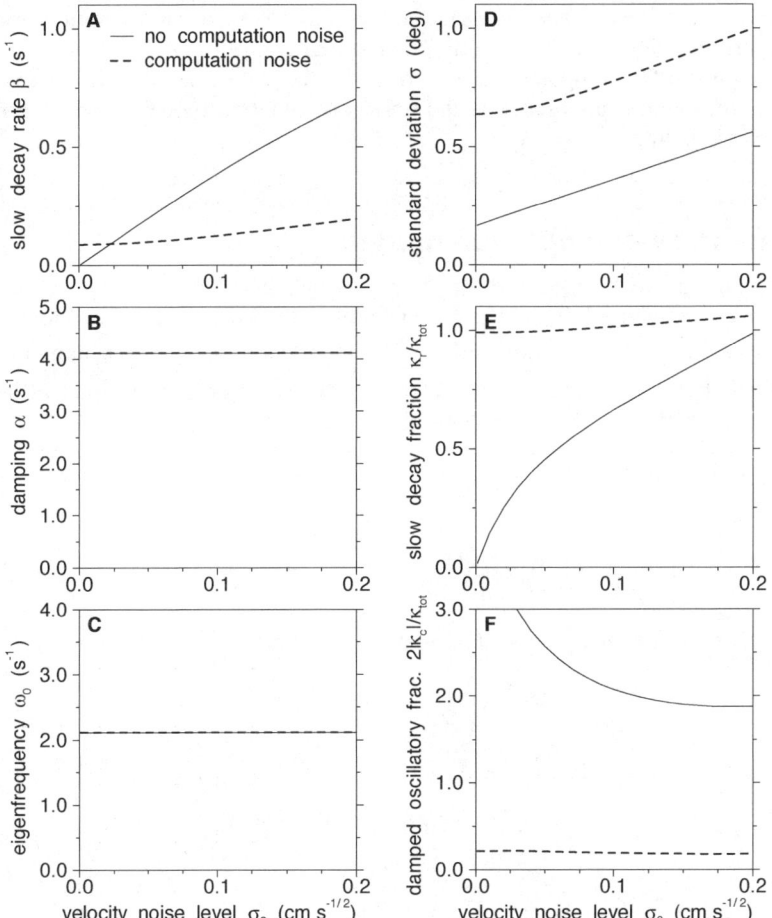

Figure 3. Average model parameters for each sensory condition. a) rate constant of the slow decay (β). b) damping (α). c) eigenfrequency (ω_0). d) COM position standard deviation (σ). e) slow-decay fraction (κ_r/κ_{tot}). f) damped-oscillatory fraction ($2|\kappa_c|/\kappa_{tot}$). Error bars denote SE of the mean.

Having identified these empirical constraints using a descriptive model, we then used them to test two simple mechanistic models of postural control: a PID-control model in which the system's state is assumed to be known (a stochastic version of Johannson et al., 1988); and an optimal control model in which the system's state is estimated by fusing multisensory information using a Kalman filter (e.g., Gusev and Semenov, 1992). Neither model was consistent with the empirical constraints identified by our descriptive model. For example, the PID-control model predicted a negative slow-decay fraction, contrary to the positive fraction near 1 found with our descriptive model. Likewise, the optimal-control model predicts a slow-decay fraction approaching zero as more sensory information is added, again contrary to the constant slow decay fraction found by our descriptive model. Consequently, the optimal control model was modified by adding noise to the computations performed by the state estimator (Kiemel et al., 2002). We refer to this model as the noisy computation model which is briefly summarized below.

7. The noisy computation model

The noisy computation model has four variables: the position x_1, the velocity x_2, the estimated position $\hat{x}_1$, and the estimated velocity $\hat{x}_2$. In the experiments described above, position x_1 is the anterior-posterior angle of the center of mass. The time derivatives of the variables are given by

$$\dot{x}_1 = x_2, \tag{2a}$$

$$\dot{x}_2 = \gamma x_2 - c_1 \hat{x}_1 - c_2 \hat{x}_2 + \sigma \xi(t), \tag{2b}$$

$$\dot{\hat{x}}_1 = \hat{x}_2 + K_{11}(z_1 - \hat{x}_1) + K_{12}(z_2 - \hat{x}_2) + K_{13}(z_3 - \gamma \hat{x}_1) \\ + \sigma_{c1} \xi_{c1}(t), \tag{2c}$$

$$\dot{\hat{x}}_2 = \gamma \hat{x}_2 - c_1 \hat{x}_1 - c_2 \hat{x}_2 + K_{21}(z_1 - \hat{x}_1) + K_{22}(z_2 - \hat{x}_2) \\ + K_{23}(z_3 - \gamma \hat{x}_1) + \sigma_{c2} \xi_{c2}(t), \tag{2d}$$

where

$$z_1 = x_1 + \sigma_1 \xi_1(t),$$
$$z_2 = \dot{x}_1 + \sigma_2 \xi_2(t) = x_2 + \sigma_2 \xi_2(t),$$
$$z_3 = \ddot{x}_1 + \sigma_3 \xi_3(t) + c_1 \hat{x}_1 + c_2 \hat{x}_2$$
$$= \gamma x_1 + \sigma \xi(t) + \sigma_3 \xi_3(t),$$

$\xi(t)$, $\xi_1(t)$, $\xi_2(t)$, $\xi_3(t)$, $\xi_{c1}(t)$ and $\xi_{c2}(t)$ are independent white-noise processes, and the K_{jk} are chosen to minimize the estimation performance index

$$J = E[d_1(x_1 - \hat{x}_1)^2 + d_2(x_2 - \hat{x}_2)^2], \quad (4)$$

where d_1 and d_2 are positive.

Equations (2a) and (2b) describe the dynamics of an inverted pendulum. The right-hand-side of (2b) consists of γx_2, the acceleration produced by gravity, and $-c_1 \hat{x}_1 - c_2 \hat{x}_2 + \sigma \xi(t)$, the acceleration produced by muscle activity, where $u(\hat{x}_1, \hat{x}_2) = -c_1 \hat{x}_1 - c_2 \hat{x}_2$ is the control function and $\sigma \xi(t)$ is process noise.

Equations (2c) and (2d) describe the dynamics of estimating position and velocity based on noisy sensory measurements; z_1 is a noisy measurement of position, z_2 is a noisy measurement of velocity, and z_3 is a noisy measurement of acceleration, transformed by subtracting the control function $u(\hat{x}_1, \hat{x}_2)$. The coefficients K_{jk} are sensory weights. They are chosen to minimize the weighted sum of squared estimation errors given by the performance index (3). The weighting of position and velocity errors does not effect the choice of sensory weights. Therefore, we set the performance-index coefficients d_1 and d_2 both equal to 1 in their respective units.

When the sensory weights K_{jk} are zero, (2c) and (2d) are an internal model of the inverted pendulum. The terms $\sigma_{c1}\xi_{c1}(t)$ and $\sigma_{c1}\xi_{c1}(t)$ describe computation noise. Computation noise is meant to model errors made by the neural systems that fuse sensory information to produce the state estimates $\hat{x}_1$ and $\hat{x}_2$. It differs from measurement noise in that it effects the dynamics of estimation even in absence of the sensory information. When the computation-noise levels σ_{c1} and σ_{c2} are zero, (2c) and (2d) are a Kalman filter (Bryson & Ho, 1975). The model has a total of 9 parameters: the inverted-pendulum parameter γ; the control-function coefficients c_1 and c_2; the process-noise level σ; the sensory-noise levels σ_1, σ_2 and σ_3; and the computation-noise levels σ_{c1} and σ_{c2}.

The autocovariance function of the model has the form

$$E[x_1(t)x_1(t+\tau)] = \kappa_{e1} e^{\lambda_{e1}|\tau|} + \kappa_{e2} e^{\lambda_{e2}|\tau|} + \kappa_{c1} e^{\lambda_{c1}|\tau|} + \kappa_{c2} e^{\lambda_{c2}|\tau|}$$

The eigenvalues λ_{e1} and λ_{e2} are called the "estimation eigenvalues"; they describe the dynamics of estimation errors and depend only on γ and the noise-level parameters. The eigenvalues λ_{c1} and λ_{c2} are called the "control-function eigenvalues"; they depend only on γ and the control-function coeffcients c_1 and c_2:

$$\lambda_{c1,2} = -c_2/2 \pm i\sqrt{c_1 - \gamma - c_2^2/4}.$$

Based on comparisons of the model to experimental data (Kiemel et al., 2002), we hypothesize that $c_1 > \gamma + c_2^2/4$ so that the control-function eigenvalues are complex-valued, corresponding to a damped-oscillation. We further hypothesize that the postural-control system under normal sensory conditions resides in a parameter regime in which the process-noise level σ, the velocity sensory-noise level σ_2 and the position computation-noise level σ_{c1} are all small. This hypothesis is stated mathematically by assuming that these parameters are of order ε, where ε is a small parameter. Then one estimation eigenvalue, λ_{e1}, is of order ε, indicating a slow rate constant; and the other estimation eigenvalue, λ_{e2}, is of order $1/\varepsilon$, indicating a fast rate constant. The control-function eigenvalues, λ_{c1} and λ_{c2} are of order 1, indicating dynamics on an intermediate time scale.

The largest coefficient of the autocovariance function is the estimation coefficient K_{e1}, which is of order ε. The control-function coefficients K_{c1} and K_{c2} are of order ε^2, and the second estimation coefficient K_{e2} is of order ε^5. Therefore, the eigenvalues of a descriptive ARMA model (see above) can be related to the eigenvalues of the mechanistic noisy-computation model: the real-valued eigenvalue λ_r corresponds to the estimation eigenvalue λ_{e1}, and the complex-valued eigenvalues λ_c and $\overline{\lambda}_c$ correspond to the control-function eigenvalues λ_{c1} and λ_{c2}. There is an analogous correspondence between the coefficients of the two models.

Figure 4 illustrates a comparison between the optimal control model with and without computation noise. All six measures shown are qualitatively consistent with our descriptive results in Figure 3 only if computation noise is added to the estimator. The crucial aspect is that computation noise affects the dynamics of the

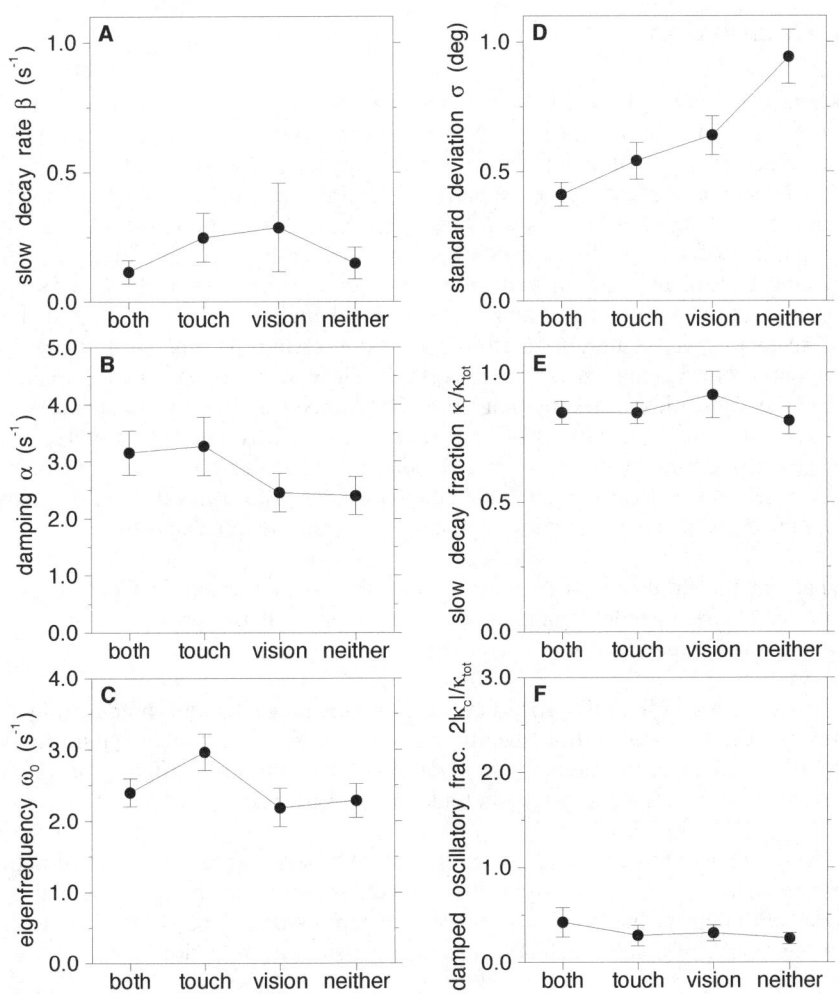

Figure 4 A–F. Measures characterizing the stochastic structure of the optimal-control model (solid lines) and noisy-computation model (dashed lines) for varying levels of velocity-measurement noise. The optimal-control model performs state estimation without computation noise. The noisy computation model includes computation noise in its state estimator and optimizes estimation rather than control. Compare to Figure 3.

estimator even in the absence of sensory information. By contrast, the effect of noise associated with the transduction of stimuli by sensory receptors (i.e., measurement noise) scale with the weights associated with those measurements.

8. Discussion

We have found that classical control theory models do not adequately account for even the relatively simple situation (i.e., biomechanical factors minimized, no perturbations, etc.) of quiet stance across varying sensory conditions. If such models cannot account for this situation, it seems unlikely that they would apply to more challenging postural situations. The noisy-computation model is a simple mechanistic model of postural control that, like optimal control models, views postural control in terms of two processes: i) the estimation of the body's state based on multisensory input and on an internal model; and ii) the use of these estimates to specify appropriate muscular responses in order to control the body's orientation. In this theoretical framework, the slow-decay eigenvalue is associated with the dynamics of estimation, and the damped-oscillatory eigenvalues are associated with the dynamics of control. Because it was able to reproduce sway properties required by our descriptive model, we argue that this is the first control-theory model that has been demonstrated to produce trajectories that are consistent with the dynamic characteristics of postural sway during quiet stance.

What are the implications of these results for postural control? Controlling an unstable inverted pendulum and the resulting variance that is present, even in quiet stance, is attributable to:

i) Process noise - the difference between the actual acceleration produced by the muscles and the acceleration specified by the controller. Multi-function muscles are involved in much more than just than the control of upright stance. The flexibility of function necessitates a trade-off with precise control.

ii) Measurement noise - any noise in the initial processing of sensory information by the nervous system to extract the relevant information (position, velocity, or acceleration) from that modality. Sensory receptors transduce stimulation with a certain degree of noise and thus reduce the precision of estimation.

Our results provide an alternative explanation. For the noisy-estimation model in our chosen parameter regime, process noise, makes only a small contribution to the variance. Most of the variance in postural sway is attributable to noise in the estimation and measurement, rather than to noise in the control process (i.e., translating the estimate of body position into a control signal). Control theory algorithms are often developed for machines (e.g., robots) in which the major sources of noise are in the sensors (measurement noise) and the plant (process noise), not the onboard computer which estimates the control signal. From a biological perspective however, significant levels of noise in a neural (central) estimator is entirely plausible. Together these findings emphasize a new source of noise whose influence can be traced to the process of fusing information from different sensory sources. This source of variation is potentially important not only for our chosen experimental system, human postural control, but any behavior which relies on more than one sensory source for successful performance.

References

Begbie JV (1967) Some problems of postural sway. In: deReuck AVS, Knight J (eds.) CIBA Foundation Symposium on myotatic, kinesthetic and vestibular mechanisms. Churchill Limited, London

Black FO, Wall C III, Nashner LM (1983) Effects of visual and support surface orientation references upon postural control in vestibular deficient subjects. Acta Oto-laryngol. 95(3-4), 199-210.

Borah JL, Young R, Curry RF (1988) Optimal estimator model for human spatial orientation. Ann NY Acad Sci 545, 51-73

Bryson AE, Ho YC (1975) Applied optimal control: Optimization, estimation, and control. Wiley, New York

Chow CC, Collins JJ (1995) Pinned polymer model of posture control. Phys Rev E 52, 907-912

Clapp S, Wing AM (1999) Light touch contribution to balance in normal bipedal stance. Exp Brain Res 125(4), 521-524

Collins JJ, DeLuca CJ (1993) Open-loop and closed-loop control of posture: A random-walk analysis of center-of-pressure trajectories. Exp Brain Res 95, 308-318

Dickstein R, Shupert CL, Horak FB (2001) Fingertip touch improves stability in patients with peripheral neuropathy. Gait Posture 14(3), 238-247

Dietz V (1992) Human neuronal control of automatic functional movements: Interaction between central programs and afferent input. Physiol Rev 72, 33-69

Dijkstra TMH, Schöner G, Giese MA, Gielen CCAM (1994) Frequency dependence of the action-perception cycle for postural control in a moving visual environment: Relative phase dynamics. Biol Cybern 71(6), 489-501

Dijkstra TMH (2000) A gentle introduction to the dynamic set-point model of human postural control during perturbed stance. Hum Movement Sci 19, 567-595

Fransson PA, Magnusson M, Johansson R (1998) Analysis of adaptation in anteroposterior dynamics of human postural control. Gait Posture 7, 64-74

Giese MA, Dijkstra TMH, Schöner G, Gielen CCAM (1996) Identification of the nonlinear state space dynamics of the action-perception cycle for visually induced postural sway. Biol Cybern 74, 427-437

Gusev V, Semenov L (1992) A model for optimal processing of multisensory information in the system for maintaining body orientation in the human. Biol Cybern 66, 407-411

Holden M, Ventura J, Lackner JR (1994) Stabilization of posture by precision contact of the index finger. J Vestib Res 4(4), 285-301

Horak FB, Macpherson JM (1996) Postural orientation and equilibrium. In: Shepard J, Rowell L (eds) Handbook of physiology. Oxford University Press, New York, 255-292

Jeka JJ, Lackner JR (1995). The role of haptic cues from rough and slippery surfaces in human postural control. Exp Brain Res. 103(2), 267-276.

Jeka JJ, Lackner JR (1994) Fingertip contact influences human postural control. Exp Brain Res 100, 495-502

Jeka JJ, Schöner G, Dijkstra TMH, Ribeiro P, Lackner JR (1997) Coupling of fingertip somatosensory information to head and body sway. Exp Brain Res 113, 475-483

Jeka J, Oie K, Schöner G, Dijkstra T, Henson E (1998) Position and velocity coupling of postural sway to somatosensory drive. J Neurophysiol. 79(4), 1661-1674.

Johansson R, Magnusson M, Akesson M (1988) Identification of human postural dynamics. IEEE T Bio-med Eng 35, 858-869

Kiemel T, Oie KS, Jeka JJ (2002) Multisensory fusion and the stochastic structure of postural sway. Biol Cybern 87(4), 262-77

Kuo AD (1995) An optimal control model for analyzing human postural balance. IEEE T Bio-med Eng 42, 87-101

Nashner LM (1981) Analysis of stance posture in humans. In: Towe A, Luschei E (eds.) Handbook of behavioral neurobiology, Vol 5, Motor coordination. Plenum Press, New York, 527-565

Nashner LM (1982) Adaptation of human movement to altered environments. TINS 5, 358-361

Newell KM, Slobounov SM, Slobounova ES, Molenaar PC (1997) Stochastic processes in postural center-of-pressure profiles. Exp Brain Res 113, 158-164

Oie KS, Kiemel T, Jeka JJ (2002) Multisensory fusion: Simultaneous re-weighting of vision and touch for the control of human posture. Cognitive Brain Res 14(1), 164-76

Peterka RJ, Benolken MS (1995) Role of somatosensory and vestibular cues in attenuating visually induced human postural sway. Exp Brain Res 105, 101-110

Peterka RJ (2000) Postural control model interpretation of stabilogram diffusion analysis. Biol Cybern 82(4), 335-343

Reginella RL, Redfern MS, Furman JM (1999) Postural sway with earth-fixed and body-referenced finger contact in young and older adults. J Vestib Res 9(2), 103-109

Riley MA, Wong S, Mitra S, Turvey MT (1997) Common effects of touch and vision on postural parameters. Exp Brain Res 117(1), 165-170

Riley MA, Balasubramanian R, Mitra S, Turvey MT (1998) Visual influences on center of pressure dynamics in upright posture. Ecol Psych 10(2), 65-91

Schöner G (1991) Dynamic theory of action-perception patterns: the "moving room" paradigm. Biol Cybern 64(6), 455-62

Teasdale N, Stelmach GE, Breunig A, Meeuwsen HJ (1991) Age differences in visual sensory integration. Exp Brain Res 85, 691-696

van der Kooij H, Jacobs R, Koopman B, van der Helm F (2001). An adaptive model of sensory integration in a dynamic environment applied to human stance control.
Biol Cybern 84(2), 103-115.

van der Kooij H, Jacobs R, Koopman B, Grootenboer H (1999) A multisensory integration model of human stance control. Biol Cybern 80, 299-308

Wolfson VJ, Whipple R, Amerman P, Kaplan J, Kleinberg A (1985) Gait and balance in the elderly. Clin Geriatr Med 1, 649-659

Woollacott MH, Shumway-Cook A, Nashner LM (1986) Aging and posture control, Changes in sensory organization and muscular coordination. Int J Aging Hum Dev 23, 97-112

Zatsiorsky VM, Duarte M (1999) Instant equilibrium point and its migration in standing tasks: Rambling and trembling components of the stabilogram. Motor Control 3, 28-38

Part IV: Perceptual and Motoric Influences on Coordination Dynamics

Governing Coordination. Why do Muscles Matter?

Richard G. Carson

Perception and Motor Systems Laboratory, School of Human Movement Studies, The University of Queensland, Brisbane, Queensland 4072, Australia.

1. Overview

The coordination of movement is governed by a coalition of constraints. The expression of these constraints ranges from the concrete – the restricted range of motion offered by the mechanical configuration of our muscles and joints; to the abstract – the difficulty that we experience in combining simple movements into complex rhythms. The diverse manner in which the constraints on coordination are expressed, belies the fact that they share common origin in the structure and organisation of the neuromuscular-skeletal system.
The balance that exists between the expression of the various constraints on sensorimotor coordination is determined by the context of the task at hand. The task context consists of elements that are intrinsic, in so much as they are related directly to the nature of the movements that are being performed; and those that are extrinsic, such as external sources of feedback. It has been customary in the literature to overlook the importance of task context in dictating the precise coalition of constraints to which a particular experimental paradigm bears witness. As a consequence, there has been a tendency to treat the various sources of constraint as distinct and exclusive (e.g. Mechsner et al., 2001), rather than recognising and understanding their cooperativity.
In the present chapter, the critical role of task context will be examined with specific reference to neuromuscular-skeletal constraints on sensorimotor coordination. In the first instance, it will be shown that both the intrinsic and extrinsic elements of the coordination task dictate the expression of constraints. It will also be demonstrated that the various constraints on coordination are complementary and inclusive, and that their expression and interaction are mediated by the integrative action of the central nervous system (CNS).

2. Task context: Intrinsic elements

2.1 Phase dependence

In a recent series of studies, we examined the proposal that the efficiency with which movements can be generated by the CNS represents a potent constraint on the stability of sensorimotor coordination. The experimental paradigm that was employed to examine this issue was very straightforward. Volunteer participants were required to make flexion and extension movements of the index finger (about the metacarpal-phalangeal joint) in time with an auditory metronome. In one condition (flex-on-the-beat), the participants were asked to synchronise the point at which the finger reaches its most flexed position with the beat of the metronome. In the other condition (extend-on-the-beat), it was the point at which the finger reached its most extended position that was to coincide with the beat of the metronome. The frequency of the metronome was increased in a series of eight steps (of eight seconds duration) over the course of just over a minute. Typically, the pacing frequency was increased in increments of 0.25 Hz, from a starting value of 2.00 Hz, to a final frequency of 3.75 Hz (e.g. Carson & Riek, 1998).

When individuals were asked to flex on the beat, they generally had little difficulty in maintaining the required pattern of coordination as the pacing frequency was increased. In marked contrast, when participants attempted to extend on the beat of the metronome, the pattern became highly unstable and deviations from the required mode of coordination were observed as the pacing frequency was increased. In many instances, the participants switched involuntarily to the flex-on-the-beat pattern of coordination (Carson, 1996; Carson & Riek, 1998). Evidently, there is a marked difference between the stability of patterns in which the flexion phase of the movement cycle is synchronised with an external signal, and those in which extension phase is the focal element of coordination.

There are a number of neuro-anatomical factors that distinguish the control of the upper limb flexors from that of the extensors. As a result of their phylogenetic origin as anti-gravity muscles, the flexors are stronger than the extensors, requiring a smaller proportion of motor units to be activated in order to produce a given level of force (Vallbo & Wessberg, 1993). Unit changes in the firing rate of corticomotoneuronal cells that innervate flexor muscles result in a greater increase of torque than an equivalent change in cells facilitating extensor muscles (Cheney, Fetz, & Mewes, 1991). In non-human primates, the proportion of motoneurons receiving monosynaptic excitation is greater for the flexor muscles of the upper arm than for the extensor muscles (Phillips & Porter, 1964). Recent studies employing transcranial magnetic stimulation of the motor cortex have provided evidence of a corresponding pattern of organisation in humans. Corticospinal neurones with direct facilitatory projections to motoneurons of elbow flexors are more numerous than neurones that project directly to elbow extensors (Palmer & Ashby, 1992). These properties are also reflected in the patterns of brain activity, which occur during functional movement tasks. Cortical activation, registered by fMRI, is substantially greater during extension than during flexion movements of the fingers, even when the relative activity of the extensor and flexor muscles is equivalent (Yue et al., 2000). The conclusion that can be drawn is that flexion movements are generated with greater efficiency than extension movements.

2.2 Neural mediation of phase dependence

It has been proposed previously that the potential for interference between functionally proximal areas of the cerebral cortex is contingent upon the degree to which these areas are activated (Kinsbourne & Hicks, 1978). It is well established that the cortical representations of muscles overlap broadly (e.g. Humphrey, 1986; Lemon 1988). In addition, it is clear that movement of any limb segment is mediated by activity that is widely distributed throughout the motor cortex (Schieber, 2001). Beta range synchronization of cortical activity and the electromyogram (EMG), in particular, is widely distributed across multiple motor and premotor areas during finger movements (Feige et al., 2000). These considerations point to the means by which interference may occur between the cortical representations of the focal muscles recruited in a movement task and of those that would be otherwise disengaged.

There is extensive evidence that the distribution of brain activity associated with a functional movement task is determined by the required level of muscle activation. The overall extent of primary motor cortex that is activated increases with the rate of movement (Blinkenberg et al., 1996; Schlaug et al., 1996). Close relationships also exist between levels of muscle activation, and fMRI measured brain activity recorded both in motor-function related cortical fields, and across the entire brain (Dai et al., 2001). With rising levels of finger flexion force, there is an initial steep increase in motor cortex activity, reflecting the recruitment of a larger number of small motor units. This is followed by a further less rapid increase in activity as a smaller number of large units are engaged (Dettmers et al., 1995).

We have argued elsewhere that the efficiency with which an action can be generated by the neuromuscular-skeletal system determines the spatial distribution of activity within the motor cortex (Carson et al., 1999). The observation that cerebral activity during movements that accentuate extension is greater than in movements in which flexion is accentuated (e.g. Yue et al., 2000), is consistent with this proposition. It is also to be supposed that these factors will impact upon the level of brain activity present during functional movement tasks. If the extent and amplitude of activity dictates the degree of interference that occurs via the overlapping cortical territories of individual muscles, there will be consequential changes in the stability of sensori-motor coordination.

2.3 Acute and chronic changes in muscle strength

We have proposed that factors that alter the efficiency with which movement is generated by the motor system, engender corresponding changes in the stability of sensorimotor coordination. We have also suggested that these effects are mediated by modifications in the activation of the higher brain centres that give rise to descending motor commands. In particular, induced changes in the efficiency with which actions are generated by the motor system, are thought to determine the level of activity in motor centres that are not engaged directly in the task. It follows, therefore, that chronic changes in muscle strength accruing from resistance training, and acute changes in force generating capacity arising from muscle damage or changes in muscle length and moment arm, will alter the efficiency of motor output, modify the distribution of activity in motor cortex, and impact upon the stability of sensorimotor coordination. These proposals have been examined in a series of studies conducted recently in our laboratory.

In the context of the finger synchronisation task described above, the frequency at which the extend-on-the-beat pattern of coordination becomes compromised is determined by the lengths of the finger extensor muscles and by their moment arms (Carson, 1996; Carson & Riek, 1998). It is well known that the capacity of a muscle to generate force is dependent upon its length (e.g. Gordon, Huxley, & Julian, 1966). Furthermore, the muscle moment arm determines the relationship between the level of motor unit recruitment and the resulting joint torque. Variations in either muscle length or moment arm therefore impact upon the level of descending drive from the higher motor centres that is required to bring about a specific movement outcome.

Chronic changes in the strength of specific muscles, such as those brought about by resistance training, also have a corresponding influence upon the stability of sensorimotor coordination. Specifically, increases in the strength of the muscles that extend the index finger enhance the stability of a syncopation task, in which individuals are required to perform a finger extension movement between the beats of an auditory metronome (Carroll et al., 2001). One direct consequence of resistance training is an increase in the gain of the corticospinal pathway (Carroll, Riek, & Carson, 2002). Following training, the level of cortical input to the spinal motoneurons that is necessary to generate a particular degree of muscle activation or joint torque is less than that required prior to training. The consequences of both chronic (Carroll et al., 2001) and acute (Carson, 1996; Carson & Riek, 1998) changes in torque output, are therefore consistent with the proposition that factors that alter the efficiency of motor output impact in a systematic fashion upon the stability of sensorimotor coordination.

In the aforementioned studies, it was not possible to determine whether the observed changes in the stability of coordination, were associated with modifications in the distribution of activity in the cortex. Converging evidence has however been provided by two related studies. In the first of these (Carson et al., 1999), volunteers were were required to perform a primary task, consisting of rhythmic flexion and extension of the index finger, while being paced by an auditory metronome, in one of two modes of coordination: flex-on-the-beat or extend-on-the-beat. Using a classical dual-task methodology, we demonstrated that the time taken to react to an unpredictable visual probe stimulus (the secondary task), by means of a pedal response, was greater when the extension phase of the finger movement was made on the beat of the metronome, than when the flexion phase was coordinated with the beat. We also manipulated the posture of the wrist in order to alter the operating lengths of muscles that flex and extend the index finger. The time required to respond to the probe stimulus was altered in a systematic fashion by this manipulation, even though the effector (foot) used to respond to the visual signal was not affected. The finding that the torque generating capacity of the muscles engaged in the primary task dictates the level of performance of the secondary task, is consistent with the view that factors that impact upon the efficiency of motor output alter the distribution of neural activity within the cortex.

In a further study, we demonstrated in a very direct fashion that interventions that diminish the force generating activity of skeletal muscle lead to increases in the spread of activation within motor cortex (Carson et al., 2002a). In this instance, acute fatigue was induced in a muscle of the upper arm (triceps brachii) by a regimen of eccentric contractions. As a result of this intervention, the force generating capacity of the muscle was reduced by approximately one third. The amplitude of the EMG activity recorded from the muscle during subsequent test contractions was substantially elevated. In addition, cortically evoked potentials recorded from a muscle (extensor carpi radialis) that was not engaged in the task, were substantially larger than those elicited prior to exercise. These observations indicate that, when the level of descending neural drive to a muscle is increased, for example to overcome the damage induced by prior eccentric contractions, there is substantial spread of excitability to other motor centres.

2.4 Bimanual coordination

What are the implications of these findings for our understanding of interlimb coordination? In the first instance, it is clear that the manner in which movements are generated by the neuromuscular-skeletal system has a profound influence on the stability of coordination. In addition, it appears that particular combinations of muscle actions are likely to be more stable than others. For example, the greater efficiency with which movement is generated by the flexors of the upper limb may lead us to supppose that the most stable patterns of bimanual coordination will be those in which flexors are active simultaneously.

In an experiment conducted to examine this possibility (Riek et al., 1992), volunteer participants were asked to perform bimanual rhythmic flexion and extension of the index fingers. The movements were performed with each forearm in one of two positions: neutral or prone. When the forearm was in a neutral position, flexion of the finger resulted in movement towards the body midline. In contrast, when the forearm was in a prone position, movement towards the midline of the body occured during the extension phase. Patterns of bimanual coordination were defined simply in terms of the spatial relationship between the index fingers.

Patterns in which the fingers were moving in the same direction were defined as inphase, and those in which the limbs were moving in the opposite direction were defined as antiphase. The particular muscle groups which were active in each pattern were determined by the relative positions of the forearms. The stability of the various patterns of coordination was invesigated using an experimental protocol (Kelso, 1984), in which the frequency of a pacing metronome was increased in a series of discrete steps.

The results were very striking. In conditions in which homologous muscles were active simultaneously, the required pattern of bimanual coordination was maintained throughout. Loss of stability occurred with increases in movement frequency only in those conditions in which generation of the target pattern of coordination required the simultaneous activation of non-homologous muscle groups (e.g. flexors in the right limb and extensors in the left limb). Transitions from the antiphase: (left arm) prone - (right arm) prone condition were universal, and three of four subjects experienced transitions from the antiphase: (left arm) neutral - (right arm) neutral condition. In addition, all of the participants exhibited transitions from the inphase condition in which the left forearm was in a prone position and the right forearm in a neutral position. Furthermore, two of four subjects demonstrated a loss of stability in the inphase condition in which the left forearm was in a neutral position and the right forearm in a prone position. In this task context therefore, the most stable patterns of bimanual coordination were those in which homologous muscles were active simultaneously. The stability of interlimb coordination was not strongly dependent upon the relative direction in which the limbs were moving.

A quite different pattern of results was obtained by Mechsner and colleagues (2001). In this instance, the primary movement task consisted of abduction and adduction of the index finger. In separate conditions, each forearm was placed in either a supine or a mid-prone position. When the forearm was supine, abduction of the finger resulted in movement away from the body midline. In contrast, when the forearm was in a mid-prone position, movement towards the midline of the body occured during the abduction phase. Patterns of coordination in which the fingers moved simultaneously towards and away from the midline of the body were termed symmetric. Patterns in which one finger moved towards the midline, while the other finger moved away from the midline, were termed parallel. When the positions of the forearms were congruent (both supine or both mid-prone), generation of the symmetric pattern of coordination required simultaneous abduction and simultaneous adduction of the two fingers. In contrast, when the positions of the forearms were incongruent, generation of the symmetric pattern of coordination required that abduction of the finger of the left hand occured simultaneously with adduction of the finger of the right hand, and vice-versa. Mechsner et al. reported that, regardless of the relative position of the forearms, the symmetric pattern was more stable than the parallel pattern of coordination. As in the incongruent positions, the production of the symmetric pattern required that the abductors of the finger of one limb were activated simultaneously with the adductors of the other limb, these data appear to suggest that muscle activation

patterns do not impose strong constraints on the stability of bimanual coordination.
What is the basis of this contrariety? The production of abduction-adduction movements of the finger can be brought about only via complex control strategies that inhibit the expression of primal muscle synergies (Carson & Riek, 2001; see also Kelso et al., 1993). In contrast, rhythmic flexion and extension of the index finger requires the alternating recruitment of muscles that have mechanically opposed lines of action. As such, this action is supported by a host of neural circuitry that promotes the reciprocal activation of anatomical (as opposed to functional) agonist and antagonist muscles.

These differences are likely to have profound implications for bimanual coordination. It is now well established that during the rhythmic alternating contraction of classically defined agonist and antagonist pairs (e.g. the flexors and extensors of the wrist), there is phasic modulation of the excitability of the homologous motor pathways of the opposite (quiescent) limb. In these circumstances, the level of excitability of the corticospinal motor pathway is positively related to the degree of recruitment of the corresponding muscle of the moving limb (Carson, Riek, & Bawa, 1999). It has also been established, however, that the degree of crossed facilitation is not simply contingent upon the level of activation of the muscles of the moving limb. It is also sensitive to the mechanical context in which the movements are performed, and by the specific muscle synergy that is thus required (Carson & Riek, 2000). In short, intrinisic features of the task, such as the means by which the focal movements are generated by the motor system, determines the nature of the neural coupling between the limbs. It is likely therefore that the crossed modulation of excitability in corticospinal motor pathways that occurs during abduction-adduction movements of the index finger, is quite distinct from that present during flexion-extension movements. As a consequence, the *relative* significance of other constraints, such as those which are expressed in the differential stability of symmetrical and parallel movements, will vary in corresponding fashion. It is proposed that this may account for the difference between the results obtained by Riek et al. (1992), and those reported by Mechsner et al. (2001).

2.5 Additional evidence of neuromuscular-skeletal constraints on interlimb coordination

The widespread presence of neuromuscular-skeletal constraints on coordination can be demonstrated by relatively subtle variations in intrinsic elements of the task context. If during pronation and supination of the forearm, the axis of rotation is adjacent to the radius, synchronisation of supination with the beat of a metronome, is more stable than a pattern in which pronation is synchronised with the metronome. In contrast, if the rotations are about an axis that is to the ulnar side of the forearm, patterns of movement requiring pronation on-the-beat are more stable than patterns requiring supination on-the-beat (Carson et al., 2000). In the context of bimanual movements, these constraints exert a powerful influence upon the dynamics of interlimb coordination (Byblow et al., 1994). When the position of the axis of rotation is equivalent for the left and the right limbs, transitions from antiphase (left limb pronating and right limb supinating simultaneously, and vice-versa) to inphase (both limbs pronating simultaneously, and supinating simultaneously) patterns of coordination are observed. In marked contrast, when the position of the axis of rotation for the left and right limb is contradistinct, transitions from inphase (symmetric) to antiphase (parallel) patterns of coordination predominate (Carson et al., 2000).

What are the mechanisms that affect such profound changes in the coupling between the limbs? Mechanical context determines the changes in muscle length and moment arm that occur during movement, and thus regulates the capacity of individual muscles to contribute to net joint torque. The central nervous system (CNS) displays a sensitivity to these factors in the composition of muscle synergies. In circumstances in which the radius can rotate freely about the ulna, flexor carpi radialis (FCR) exhibits a moment arm for pronation, that is of a magnitude similar to those observed for pronator teres and pronator quadratus (Ettema et al., 1998). During movements executed in this mechanical context, FCR is activated strongly as the forearm moves from supination into pronation. In contrast, when the axis of rotation is displaced towards the radius the muscle is disengaged (Carson et al., 2000). As we have seen previously, alterations in the composition of muscle synergies impact directly upon the neural coupling between the limbs (Carson & Riek, 2000).

While there are specific neural structures that provide the substrate for the phenomenology of bimanual coordination, many of the prototypical features are also exhibited prominently in movements involving the coordination of the upper and lower limbs (e.g. Baldissera et al., 1982; Carson et al., 1995; Kelso & Jeka, 1992). Clearly, the *structures* implicated in the coupling of the hands are not necessarily those that mediate interactions between the arm and the leg. The evidence that is presently available suggests, however, that the neural *mechanisms* are similar in nature. In particular, the pattern of recruitment of the muscles of the lower limb impacts upon the excitability of the motor centres that are enagaged during movement of the upper limb (Baldissera et al., 2002). Furthermore, the nature of the neural coupling appears to be that which would favour isodirectional movements of the upper and lower limbs, which are known to be more stable than non isodirectional movements (Swinnen, 2002).

3. Task context: Extrinsic elements

3.1 A coalition of constraints

Clearly there are a host of neuromuscular-skeletal constraints on interlimb coordination. The degree to which they are exhibited depends in large measure on intrinsic elements of the task context. Specifically, their expression is related directly to the means by which the focal movements are generated by the motor system. The composition of the muscle synergies required by the task also determines the nature of the neural coupling between the limbs. In so much as the of neuromuscular constraints on coordination are altered in a task dependent fashion, it follows that the *relative* significance of other classes of constraint will vary in a corresponding manner. It should be evident from the foregoing that the various elements of constraint on coordination need not be conceived of as being mutually exclusive. Rather, the balance of their expression is influenced in a systematic manner by the intrinsic and extrinsic elements of the task.

We recently conducted an experiment in which the extrinsic elements of a sensorimotor coordination task were altered through the provision of augmented feedback in another modality (Kelso et al., 2001; see also Byblow et al., 1995). The experimental paradigm was a variant of the synchonisation task that has been described previously. A group of volunteers produced rhythmic flexion and extension movements of the index finger, in time with the beat of an auditory metronome. The movements were made in either a flex-on-the-the-beat or an extend-on-the-beat pattern of coordination. When the participants produced the extend-on-the-beat pattern, there was a tendency to switch to flex-on-the-beat as the frequency of the pacing metronome was increased (e.g. Carson, 1996). In some instances, we also required that the participants make contact with a physical stop, the location of which was either coincident with or counterphase to the auditory stimulus. Two effects were observed. When the haptic contact was coincident with the beat, both patterns of coordination (flex-on-the-beat and extend-on-the-beat) were stabilised. In contrast, when the haptic contact was counterphase to the metronome, coordination was actually destabilised, with transitions occurring from flex-on-the-beat to extend-on-the-beat and vice-versa. Evidently individuals are drawn to patterns of coordination in which sound and touch coincide in time, regardless of whether flexion or extension movements are accentuated. These results indicate that, by changing the extrinsic elements of the task, the coalition of constraints that governs sensorimotor coordination can be varied in a systematic fashion.

3.2 Augmented feedback and bimanual coordination

It is also well known that the provision of augmented feedback can have a profound bearing on the stability of bimanual coordination (e.g. Byblow et al., 1999). Indeed, this has been used in a very practical way to render tractable the study of patterns of interlimb coordination that cannot otherwise be produced in a stable fashion. It often takes considerable practice to perform a 90 degrees out of phase pattern of coordination, in which one limb leads the other by a quarter of a movement cycle. In order to expedite this process, it has become custumary to provide continuous visual feedback of the relative phase relationship between the hands in the form of a Lissajous plot (Lee et al., 1995). This consists of a real-time display of the angular displacement of the right limb (e.g. the ordinate) plotted against the angular displacement of the left limb (e.g. the abscissa). If a relative phase relation of 90 degrees is maintained, and the motion generated by the two limbs is of the same amplitude, a perfectly circular trace will be described. In the event that the displacements of the limbs are not equivalent, variations from the required temporal relationship appear as deviations from an elliptical trajectory (e.g. Carson et al., 2002b). In the absence of visual feedback of this form, or explicit timing cues for the movement of each limb (e.g. Yamanishi et al., 1980), most people are unable to sustain anything other than inphase or antiphase patterns of coordination. The observation that the provision of meaningful augmented feedback in the form of a Lissajous plot releases this restriction, provides further substance to the proposal that varying the extrinsic elements of the task alters the expression of constraints on coordination.

The production of non-integer frequency relationships between the movements of the two hands (e.g. 4:3) can remain intractable to all but the most accomplished performers (Summers et al., 1993). It has been shown recently, however, that through the provision of suitably structured augmented feedback, otherwise unstable patterns of interlimb coordination can be generated in a consistent and reliable fashion (Mechsner et al., 2001). When synchronous (i.e. 1:1) clockwise (or anticlockwise) circling movements are required of both arms, the instructed pattern of coordination becomes virtually impossible to sustain as the frequency of movement is increased (Semjen et al., 1995). One would anticipate therefore that it would be exceptionally difficult to produce bimanual circling movements in a 4:3 frequency ratio. Yet it seems that if the extrinsic elements of the task are configured appropriately, this need not be the case. In a unique variation of the bimanual circling paradigm, Mechsner and colleagues used a gear system to manipulate the relationship between circular motions of the arms, and visual signals that were used as proxies for the positions of the hands. As a result of the gearing, an instructed 1:1 frequency relationship between the visual proxies could be brought about only by motions of the arms that were in a 4:3 frequency ratio. On this basis, the required frequency relationship between the limbs could be generated successfully in two symmetrical configurations, in which one proxy moved clockwise while the other moved counterclockwise (the most stable pattern for conventional bimanual circling movements). These results provide further confirmation that, by changing the extrinsic elements of the task in an appropriate fashion, and by providing suitably structured augmented visual feedback, otherwise unstable patterns of coordination can be maintained successfully.

4. Conclusions

The focus of the present chapter has been upon the coalition of constraints that governs human interlimb coordination. It has been proposed that the various sources of constraint should be conceived of as being inclusive and complementary, rather than exclusive and distinct. This advocacy is borne of the recognition that the mutual interplay of constraints is mediated by the integrative action of the CNS.

What are the specific constraints that govern interlimb coordination? The answer depends upon the nature of the questions that are asked of the motor system. The expression of constraints on coordination is contingent upon the experimental model that is employed, and the particular task manipulations that are applied. In this regard, it is essential to consider both the intrinsic and extrinsic elements of the task context. The intrinsic elements of a task, are those that are related directly to the nature of the movements being performed. External elements include additional features such as the provision of augmented feedback. In the present chapter it has been shown that both the internal and external elements of coordination tasks, alter the expression of constraints in profound and systematic ways. The conclusion thus drawn is that the various constraints on coordination are complementary, rather than mutually exclusive.

Author Notes

This work was supported by the Australian Research Council, and by a Visiting Fellowship awarded to the author by the Katholieke Universiteit Leuven. The contribution of Stephan Riek to much of the experimental work described in this chapter is gratefully acknowledged. I also extend my thanks to Winston Byblow for stimulating discussion, and constructive commentary.

References

Baldissera F, Cavallari P, Civaschi P (1982) Preferential coupling between voluntary movements of ipsilateral limbs. Neurosci Lett 34, 95-100

Baldissera F, Borroni P, Cavallari P, Cerri G (2002) Excitability changes in human corticospinal projections to forearm muscles during voluntary movement of ipsilateral foot. J Physiol-London 539, 903-911

Blinkenberg M, Bonde C, Holm S, et al (1996) Rate dependence of regional cerebral activation during performance of a repetitive motor task: A PET study. J Cerebr Blood F Met 16, 794-803

Byblow WD, Carson RG, Goodman D (1994) Expressions of asymmetries and anchoring in bimanual coordination. Hum Movement Sci 13, 3-28

Byblow WD, Chua R, Goodman D (1995) Asymmetries in coupling dynamics of perception and action. J Motor Behav 27, 123-137

Byblow WD, Chua R, Bysouth-Young DF, Summers JJ (1999) Stabilisation of bimanual coordination through visual coupling. Hum Movement Sci 18, 281-305

Carroll TJ, Barry B, Riek S, Carson RG (2001) Resistance training enhances the stability of sensori-motor coordination. Proc R Soc Lond B 268, 221-227

Carroll TJ, Riek S, Carson RG (2002) The sites of neural adaptation induced by resistance training in humans. J Physiol-London 544, 641-652

Carson RG (1996) Neuromuscular-skeletal constraints upon the dynamics of perception-action coupling. Exp Brain Res 110, 99-110

Carson RG, Riek S (1998) The influence of joint position on the dynamics of perception-action coupling. Exp Brain Res 121, 103-114

Carson RG, Riek S (2000) Musculo-skeletal constraints on corticospinal input to upper limb motoneurones during coordinated movements. Hum Movement Sci 19, 451-474

Carson RG, Riek S (2001) Changes in muscle recruitment patterns during skill acquisition. Exp Brain Res 138, 71-87

Carson RG, Riek S, Bawa P (1999) Electromyographic activity, H-reflex modulation, and corticospinal input to forearm motoneurones during active and passive rhythmic movements. Hum Movement Sci 18, 307-343

Carson RG, Riek S, Smethurst CJ, Lison JF, Byblow WD (2000) Neuromuscular-skeletal constraints upon the dynamics of unimanual and bimanual coordination. Exp Brain Res 131, 196-214

Carson RG, Chua R, Byblow WD, Poon P, Smethurst CJ (1999) Changes in posture alter the attentional demands of voluntary movement. Proc R Soc Lond B 266, 853-857

Carson RG, Goodman D, Kelso JAS, Elliott D (1995) Phase transitions and critical fluctuations in rhythmic coordination of ipsilateral hand and foot. J Motor Behav 27, 211-224

Carson RG, Riek S, Shahbazpour, N (2002a) Central and peripheral mediation of force sensation following eccentric or concentric contractions. J Physiol-London 539, 913-925

Carson RG, Smethurst CJ, Forner M, Meichenbaum, DP, Mackey DC (2002b) The role of peripheral afference during acquisition of a complex coordination task. Exp Brain Res 144, 496-505

Cheney PD, Fetz EE, Mewes K (1991) Neural mechanisms underlying corticospinal and rubrospinal control of limb movements. Prog Brain Res 87, 213-252

Dai TH, Liu JZ, Sahgal V, Brown RW, Yue GH (2001) Relationship between muscle output and functional MRI-measured brain activation. Exp Brain Res 140, 290-300

Dettmers C, Fink GR, Lemon RN, Klaus MS, et al (1995) Relation between cerebral activity and force in the motor areas of the human brain. J Neurophysiol 74, 802-815

Ettema GJC, Styles G, Kippers V (1998) The moment arms of 23 muscle segments of the upper limb with varying elbow and forearm positions: Implications for motor control. Hum Movement Sci 17, 201-220

Feige B, Aertsen A, Kristeva-Feige R (2000) Dynamic synchronization between multiple cortical motor areas and muscle activity in phasic voluntary movements. J Neurophysiol 84, 2622-2629

Gordon AM, Huxley AF, Julian FJ (1966) The variation in isometric tension with sacromere length in vertebrate muscles. J Physiol-London 184, 170-192

Humphrey DR (1986) Representation of movements and muscles within the primate precentral motor cortex: historical and current perspectives. Faseb J 45, 2687-2699

Kelso JAS (1984) Phase transitions and critical behavior in human bimanual coordination. Am J Physiol 240, R1000-1004

Kelso JAS, Buchanan JJ, DeGuzman GC, Ding M (1993) Spontaneous recruitment and annihilation of degrees of freedom in biological coordination. Phys Lett A 179, 364-371

Kelso JAS, Jeka, JJ (1992) Dynamic patterns and direction-specific phase transitions in human multi-limb coordination. J Exp Psychol Human 18, 645-668

Kelso JAS, Fink PW, DeLaplain CR, Carson RG (2001) Haptic information stabilizes and destabilizes coordinated movement. Proc R Soc Lond B 268, 1207-1213.

Kinsbourne M, Hicks RE (1978) Mapping functional cerebral space: competition and collaboration in human performance. In: Kinsbourne M (ed.) Asymmetrical Function of the Brain. Cambridge University Press, 267-273

Lee TD, Swinnen SP, Verschueren S (1995). Relative phase alterations during bimanual skill acquisition. J Motor Behav 27, 263-274

Lemon R (1988) The output map of the primate motor cortex. Trends Neurosci 11, 501-506

Mechsner F, Kerzel D, Knoblich G, Prinz W (2001) Perceptual basis of bimanual coordination. Nature 414, 69-73

Palmer E, Ashby P (1992). Corticospinal projections to upper limb motoneurons in humans. J Physiol-London 448, 397-412

Riek S, Carson RG, Byblow WD (1992) Spatial and muscular dependencies in bimanual coordination. J Hum Movement Stud 23, 251-265

Schieber MH (2001) Constraints on somatotopic organization in the primary motor cortex. J Neurophysiol 86, 2125-2143

Schlaug G, Sanes JN, Thangaraj V, Darby DG, Jäncke L, Edelman RR, Warach S (1996) Cerebral activation covaries with movement rate. Neuroreport 7, 879-883

Semjen A, Summers JJ, Cattaert D (1995) Hand coordination in bimanual circle drawing. J Exp Psychol Human 21, 1139-1157

Summers JJ, Ford SK, Todd JA (1993) Practice effects on the coordination of the 2 hands in a bimanual tapping task. Hum Movement Sci 12, 111-133

Swinnen SP (2002) Intermanual coordination: From behavioural principles to neural-network interactions. Nat Rev Neurosci 3, 350-361

Vallbo ÅB, Wessberg J (1993) Organization of motor output in slow finger movements in man. J Physiol-London 469, 673-691

Yamanishi J, Kawato M, Suzuki R (1980) Two coupled oscillators as a model for the coordinated finger tapping by both hands. Biol Cybern 37, 219-225

Yue GH, Liu JZ, Siemionow V, et al (2000) Brain activation during human finger extension and flexion movements. Brain Res 856, 291-300

Guiding Movements without Redundancy Problems

Ramesh Balasubramaniam[1] & Anatol G. Feldman[2]

[1]Behavioural Brain Sciences Centre, University of Birmingham, UK.
[2]Neurological Sciences Research Centre, University of Montréal, CANADA.

Approaches to the problems of multi-muscle and joint redundancy have typically been based on the assumption that control levels of the nervous system directly deal with variables describing the motor output – electromyographic (EMG) signals, forces and kinematics. An alternative approach to these problems can be developed in the framework of the ? model based on the empirical solution of another classical problem in the motor control - that of the relationship between posture and movement. This solution implies that control levels guide movement by changing specific neurophysiological parameters and modify their pattern if the resulting action is in error. Specifically, these control parameters interfere with the transmission of afferent signals by spinal and supraspinal neurons to motoneurons. Some parameters reset the spatial coordinates at which a stable posture of the body or its segments can be reached. Other parameters deal with stability of posture and movement. This parametric control strategy releases higher control levels from the burden of solving redundancy problems at the level of output, i.e. mechanical and EMG variables. In response to changes in control parameters, appropriate values of mechanical and EMG variables and their transformations (e.g., from the hand kinematics to joint angles) emerge automatically, following the natural tendency of the neuromuscular system to reach an equilibrium state. This process results from the natural tendency to minimize the overall activity and the interactions between different components (neurons, muscles and joints) of the neuromuscular system in response to resetting of control parameters (the principle of minimal interaction). Based on these ideas, we outline non-computational, dynamical solutions of the problems of multi-muscle and multi-joint redundancy. This approach does not reject the notion of synergies, primitives, or recently proposed classification of multi-joint co-ordinations into a controlled and uncontrolled manifolds. Rather, it suggests that synergies or manifolds, like trajectories and forces, may be an emergent property of the neuromuscular behavior resulting from the response of the system to changes in control parameters in specific environmental conditions.

1. Introduction

The number of available degrees of freedom (DFs) of the body is typically greater than that required to reach the motor goal (*DFs- redundancy*). The number of muscles per one DF is much greater than 2 (*multi-muscle redundancy*). The nervous system takes advantage of these redundancies to control actions in a flexible way so that, for example, the same motor goal can be reached differently depending on our intentions, external environmental (e.g. obstacles) or intrinsic (neural) constraints. Despite this flexibility, the central control of actions is unambiguous: each time the body moves, a unique action is produced despite the possibility of using other actions leading to the same goal. It is unclear how these seemingly opposite aspects – flexibility and uniqueness- are combined in the control of actions. Following Bernstein (1967), we refer to these aspects of action production as the "redundancy problem".

Approaches to redundancy problems have typically been based on the assumption that control levels of the nervous system are directly involved in programming, computation and specification of the motor output (muscle activations, forces, and kinematics). In the most explicit form, this assumption underlies force-control (FC) models of movement production (e.g. Wolpert, Ghahramani & Flanagan, 2000). Initially stemming from robotics (Hollerbach, 1972), such models have became especially popular following the recent advances in computational neuroscience. FC models have also integrated the notion that, in movement production, the nervous system relies substantially on motor memory, takes advantage of previous experiences, learning, implicit or explicit knowledge of physical properties of the body and the environment (Lashley, 1951). In the force control formulation, this capacity has become associated with inverse and forward internal models that represent the basic dynamical properties of the system or its components. A number of recent articles have presented arguments in support of the idea that control is based on paired inverse and forward internal models (Bhushan and Shadmehr, 1999, Wolpert and Kawato, 1998; Kawato, 1999; Wolpert and Ghahramani, 2000).

FC models, however, seem cannot be reconciled with some basic physiological principles of movement production (Ostry and Feldman, 2003). A major problem with these models is their inability to explain in a physiologically feasible way how control systems produce movements without provoking resistance of posture-stabilizing mechanisms to deviations from the initial posture. This question is the essence of the posture-movement problem that in most explicit way has been formulated by Von Holst and Mittelstaedt (1950/1970). Specifically, they emphasized that there are neuromuscular mechanisms that generate electomyographic (EMG) activity and forces in order to resist perturbations that tend to deflect the body from an initial posture. But at the same time, the organism can intentionally adopt different postures and maintain them with a certain degree of stability. If each new posture of the system is considered to be a deflection from an initial posture, then the deflection would result in resistance that would tend to return the system to its initial posture. However, everyday experience tells us that the nervous system has no problem in moving the body or its segments away from an initial posture and stabilizing different postures without triggering resistance (Balasubramaniam & Wing, 2002).

A solution to the posture-movement problems could be obtained if posture stabilizing mechanisms ("reflexes") were suppressed, completely or partially, by a central pattern generator when the transition to a new posture is made. This solution has been rejected by Von Holst and Mittelstaedt 1950/1970) on the basis of empirical evidence showing that any posture of the body is maintained by resisting reactions similar to those seen in the restoration of initial posture in response to perturbations (Von Holst & Mittelstaedt, 1950/73). Von Holst and Mittelstaedt suggested that in order to produce movement, the nervous system influences afferent systems in such a way as to reset the initial postural state. As a result, the initial posture appears to be a deviation from the newly specified posture. This implies that the same postural mechanisms that stabilized the initial posture would act to stabilize the new one. The notions that postural resetting underlies movement production and that such resetting is achieved by means of appropriate central influences on afferent systems are the essence of Von Holst's reafference principle that he considered as a solution to the "posture-movement problem". Note that while formulating the reafference principle, Von Holst was well aware of the existence of central pattern generators of different movements and that active movement production is possible in the absence of proprioceptive feedback to muscles (deafferentation). This knowledge did not discourage him from the formulation of the reafference principle that implies that, in intact neuromuscular systems, proprioceptive feedback is fundamental for adequate regulation of posture and movement. We think that this implication is well justified by the results of numerous comparisons of movement production in intact and deafferented organisms, both in humans and animals. In particular, in the absence of vision, deafferented patients, even after years of practice are unable to maintain steady-state positions of the arm, walk or stand without assistance, thus implicating a strong role of proprioceptive feedback in the stability and regulation of any posture (Levin et al, 1995).

The posture and movement problem is especially obvious in FC models that are based on the idea of programming EMG signals and muscle forces (torques) according to the desired trajectory of the effector. Specifically, the programmed generation of forces will produce deviation of system from the initial position. In response, the posture-stabilizing mechanisms will produce resistance tending to bring the system to its initial position. Control levels might attempt to reinforce the programmed action by generating additional EMG and torques that counteract this resistance. However, this strategy would be non-optimal in terms of energy output, since it would require high forces not only for motion but for the maintenance of final posture. The problem is that the force control model has no means to reset the "postural state" in a physiologically feasible way. Moreover, the FC formulation conflicts with a basic physical rule stating that steady states of a dynamical system are determined by intrinsic system parameters that are independent of variables describing its external, output behavior (Glansdorff & Prigogine, 1971). Applying this rule to biological systems, one can say that transitions from one steady state body posture to another result from resetting intrinsic system parameters to which external variables (in our case EMG signals, forces and kinematics) do not belong. The output variables emerge in response to this resetting. Focusing on output variables (EMG and forces) FC models were unable to indicate specific neurophysiological parameters responsible for postural resetting. Therefore, the transition between postures remains an unresolved issue in the FC model. This is not an inconsequential failing of the force control formulation. The unopposed shift between postural states is a fundamental characteristic of everyday motor activity. The inability of the force control strategy to re-establish posture at a new location without self-generated resistance is a basic failing of the formulation. Although the force control hypothesis has important applications in robotics (Balasubramaniam & Feldman, 2002), it falls short of explaining a very basic aspect of biological movement production.

The λ model for motor control, a version of the equilibrium-point (EP) hypothesis (Feldman & Levin, 1995) fully respects the "reafference principle". In addition, the λ model explicitly indicates the parameters that the control systems can reset to produce an unopposed movement to a new posture. In order to explain how this model solves the posture-movement problem, as well as the problems of muscle and joint redundancy, we will first outline the basic features of the model.

2. Fundamentals of the equilibrium point hypothesis (the λ model)

The λ model is based on the experimental finding that central changes in length-dimensional parameters - muscle activation thresholds (also called the stretch reflex thresholds) – underlie active motor behaviors (postural resetting, voluntary movement or isometric torque generation; Matthews 1959; Asatryan and Feldman 1965; Feldman 1996; Feldman and Orlovsky 1972). Many, including cortico-spinal descending systems have the capacity to regulate muscle activation thresholds (Feldman & Orlovsky 1972). Deficits in the central regulation of thresholds are associated with dramatic movement problems in neurological patients including those with impairments of cortical areas or deafferented patients (Levin and Dimov 1997; Levin et al. 2000). The basic principles of the equilibrium point (EP) hypothesis are summarized in Figure 1.

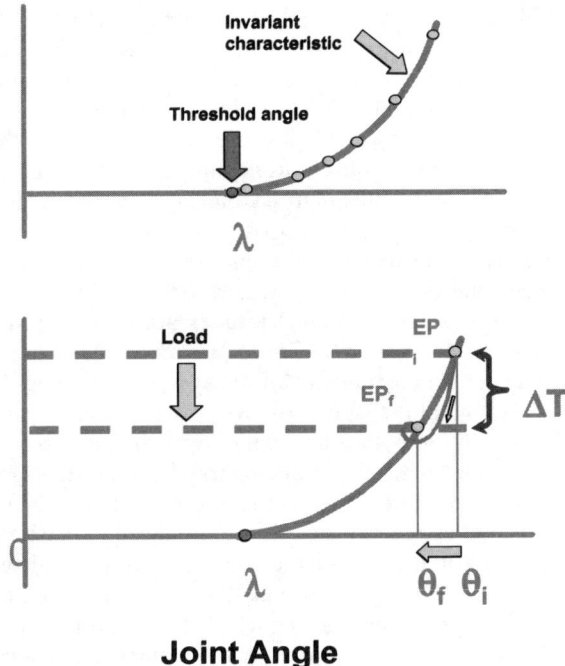

Figure 1. Fundamentals of EP hypothesis. Upper panel. Neural control levels set the threshold muscle angle (λ). Muscle activation and recruitment of its motor units change depending on the difference between the actual and the threshold joint angles. With the threshold constant, the torque-angle relationship is called the invariant characteristic, IC. Given a specific value of the threshold and external load, the system achieve a steady state in which the load is balanced. This state is characterised by the appropriate combination of position and muscle torque. This combination is called the equilibrium point or EP (the point of intersection of the load characteristic and IC). Voluntary movements are produced by setting a new value of the threshold. Lower panel. Involuntary movement. A change in the load (vertical arrow) brakes the balance between the muscle and load torques (ΔT) resulting in motion (green curve) to a new equilibrium point on the same IC (EP_f), if the subject does not intervene, at which the balance is regained.

According to this scheme, motoneurons are recruited in the range where the actual muscle length exceeds its threshold length (λ). In other words, by specifying the activation threshold, control levels "tell" motoneurons, in a feedforward manner, where, in terms of spatial coordinates, they will be recruited and counteract external forces (in particular, those of antagonist muscles). Neither the number nor the frequency of recruited motor units is specified by the control signal. Thus this model illustrates an important notion that the relationship between the central control signals and the motor output is *ambiguous* (Bernstein 1967). For example, during changes in the arm position following unloading (the unloading reflex), control signals remains the same but the motor output (EMG activity, torque, position,) varies depending on the load; in the waiter reflex (waiter holds a steady tray as he unloads dishes from it), the same position is maintained due to a change in the central control signals. Note that by specifying a central control pattern the nervous system limits the set of possible actions whereas the interaction of the neuromuscular system with the environment (the load in this case) reduces this set to a unique action. Having this in mind, one can say that central control signals are *task-specific*, i.e. produce the desired motor output only in specific external conditions. A change in these conditions may necessitate readjustment of control signals to reach the desired goal. Another important point of the model is the notion that nervous system cannot and does not need to program muscle forces, kinematics and EMG patterns. These are characteristics *emerging* from the interaction between all components of the system, including the environment. Yet another important point is that control systems take *advantage of reflexes* (e.g., by changing the stretch reflex threshold) in movement and force or torque production, which is consistent with the contemporary notion that reflexes are broadly adjustable rather than rigid stimulus-response structures. The most important point of the model is that active movements are produced by changing control variables shifting the equilibrium point of the system. This point determines the postural state (steady state) of the system and thus pre-determines the values of output variables (EMG, forces, torques) at which the neuromuscular system might be stabilized. In this sense, posture and movement are controlled by the same mechanisms, thus offering a direct solution of the Von Holst's posture-movement problem described above. Note that the basic notions described in this section are straightforward consequences of a simple experimental fact that intentional movements result from shifts in the muscle activation thresholds (Asatryan & Feldman 1965).

To summarize, the λ model suggests a physical rule for movement production given that systems with position-dependent forces have an attractor or equilibrium point. The location of this attractor in spatial coordinates is determined not by forces but by specific parameters of the systems - determinants of the equilibrium state. Since equilibrium postures of the body are not determined by muscle forces, a voluntary movement of the body from one static posture to another can only be accomplished by altering the determinants of the equilibrium state. If the nervous system could alter muscle forces leaving these determinants unchanged, muscles of the body would generate additional forces resisting the deflection from the initial posture and thus preventing the transition to a new posture. In contrast, in response to a change in these determinants, muscle activation and forces would *emerge automatically*, which makes programming or intentional specification of muscle forces unnecessary.

A natural extension of the λ model has been made on the basis of the principle of minimal interaction (Gelfand & Tsetlin, 1971). Activity of each component of the neuromuscular system depends on the difference between the actual and the referent (threshold) value of appropriate physical variable. Intentional movements or isometric torques are produced by shifting the referent values of physical variables. One can say that biological motor actions obey the principle of minimal interaction if the afferent feedback to each element and the interactions between different elements of the system are specifically organized to drive the neuromuscular system to a state in which the difference between the referent and the physical values of output variables and thus the overall activity in the system becomes minimal, in the limits determined by task constraints. The muscle contraction, stretch reflex and reciprocal inhibition between agonist and antagonist muscles are examples of the processes realizing the principle of minimal interaction. More specifically, in response to activation, muscles shorten thus decreasing the difference between the actual and the referent (threshold length) of the muscle and allowing homonymous and heteronymous proprioceptive feedback to diminish the activity (see Figure 2).

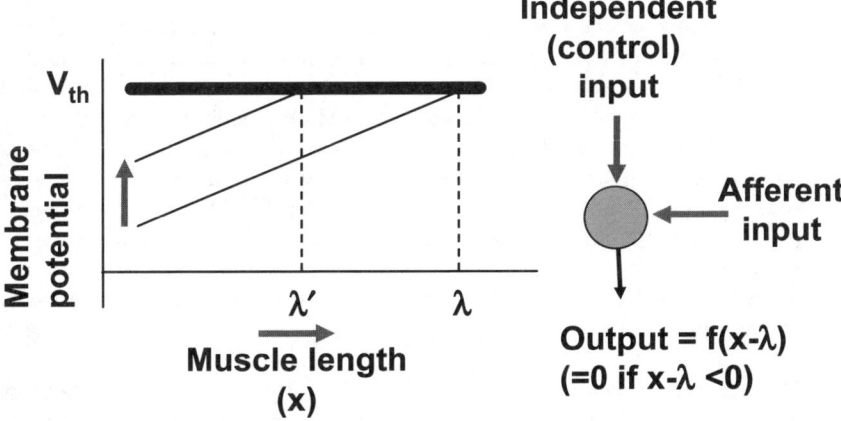

Figure 2. Frame of reference for a single motoneuron. Motoneuron together with length-dependent afferent input and electrical threshold (Vth) is a device that transforms independent, i.e. control changes in the membrane potential (red arrows) into length dimensional variable, λ. Left panel. Initially, stretching the muscle quasi-statically results in a gradual increase in the membrane potential of the motoneuron and its recruitment at the threshold length, λ. When an independent control signal is added, the motoneuron is recruited at a shorter muscle length. Thus, the control signal only set the range ($x-\lambda > 0$) of the muscle lengths, x, in which the motoneuron can be active) and counteract to external forces. Right panel. Each neuron has its own FR in which afferent input carries signals on appropriate physical variable, s (which, for a motoneurons, coincides with the muscle length, x). At the level of the membrane of the neuron, an independent control input, is transformed into a threshold value, h (coinciding with l for the motoneuron) of the same physical variable. This threshold plays of the role of the referent point for the measurement for variables. The output of the neurones is thus frame-dependent.

Although there is a tendency in literature to consider the λ model as analogous to a servo mechanism in which the controller directly specifies a desired position, the analogy is actually misleading. In the model, λ is not the desired position that the system should establish and the difference between the actual muscle length and λ is not the movement error. A more appropriate interpretation of the model is the following. There exists a spatial frame of reference (FR) or a system of coordinates in which motoneurons function. This frame has a referent or origin point - the threshold muscle length. Control systems may shift the referent point to produce muscle activation, force and movement. The neuronal activity in this frame is defined by the distance between the point characterizing the current state of the system and the origin point. A motoneuron that is recruited later in the sequence has a spatial frame embedded in the frames of motoneurons recruited earlier yielding a natural hierarchical structure for movement production. The notion that the activity of all elements of the neuromuscular systems are frame-dependent, that intentional movements are produced by shifting appropriate FRs by changing their referent parameters (origin, scales, orientation, and even geometry), that there is a hierarchical relationship between different FRs, and that all the interaction in the system, both within each FR and between different FRs, obey the principle of minimal interaction seems most useful in our approach to the problem of multi-muscle and multi- DF redundancy as well as to other problems of motor control in general. We will specifically focus on two FRs. One FR comprises all possible configurations of the body controlled by shifts in the FR origin that represents a particular, referent body configuration specified by the nervous system. The other FR is a FR associated with the environment and controlled by shifts in its origin. Like the FR for motoneurons, each FR is organized based on appropriate afferent inputs to neurons. For example, the FR associated with the environment is supposedly organized predominantly based on vestibular and visual signals.

In the subsequent sections, we will present empirical evidence that appropriate frames are actually used to guide movements in the wake of multi-muscle and multi DF redundancy.

3. Issues of redundancy

Although our solution of the DF redundancy problem has not been elaborated in detail, the main idea is simple and may be illustrated for arm pointing movements. Let us assume that some spinal and supraspinal neurons projecting to motoneurons of skeletal muscles of the body, including the extremities may integrate proprioceptive signals from muscles, joint and skin to receive afferent signals, say, about the coordinates of the tip of the index finger (the endpoint) that is typically used to point to targets. The role of these signals will be similar to those of afferents (muscle spindles) that are sensitive to changes in the muscle length, except that the recruitment and activity of these neurons will depend not on muscle length but from coordinates of the endpoint. Like for motoneurons, control influences on these neurons can be measured by the amount of shifts in the threshold (referent) coordinates of the endpoint. The difference between the actual and the referent coordinates will determine whether or not such neurons are recruited. These referent coordinates may be shifted by control levels in a FR associated with the environment to produce *a referent trajectory*. The neuromuscular system will tend to minimize the discrepancy between the actual and referent coordinates forcing the arm and other body segments to move until the endpoint reaches a final position at which a minimum of activity of the neurons and in the system in general is reached. After this, the system may compare the output with the desired one. In particular, if the final position of the arm endpoint is different from the desired one, control levels may adjust the referent endpoint trajectory until the final endpoint position coincides with the desired one. Again, although the set of possible configurations for each position of the endpoint is redundant, the minimization process initiated by shifts in the referent coordinates of the endpoint will result in a unique pattern of them. This configuration pattern can indeed, vary with repetitions, intentional modifications of the referent pattern, task constraints (including release or restriction in motion of some DFs), and history-dependent changes in the neuromuscular system (e.g. due to fatigue).

4. Multi-muscle control and the referent configuration (R) hypothesis

The configuration of the body may be described geometrically by a set of the current joint angles or coordinates q_i, associated with the n mechanical DFs of the body. These angles comprise an n dimensional vector, Q. It is assumed that the nervous system compares these values with referent angular values r_i that taken together comprises the referent configuration of the body, R. These referent angles r_i resemble the reciprocal commands previously defined in the lambda model. The model has also introduced the notion of coactivation commands (ci) that determine the extent of the spatial zone in which agonists and antagonists may be active simultaneously (Levin & Dimov, 1997). When $c_i = 0$, then the r_i command is the threshold angle at which the recruitment of muscles in the corresponding DF begins. The ensemble of these threshold angles, a multidimensional vector defines the threshold configuration of the body- R, a collective threshold for all skeletal muscles. One can say that due to the collective threshold, all muscles are involved in a synergy. The threshold nature of the R implies that when configurations coincide (Q = R or Q − R = 0), the EMG activity of the all the muscles involved should be zero independent of their biomechanical functions. Such cases are rare (but not impossible, see below) and typically the body moves until the external and muscle forces reach a balance at an "equilibrium body configuration", if stability requirements are met. In order to produce a voluntary movement, the nervous system modifies R, a change in which elicits an imbalance of forces at the initial body configuration, resulting in motion to a new configuration at which balance and stability are regained.

A slightly different scenario may result when the co-activation command is employed. This command produces changes in muscle activation thresholds to surround the R configuration with a zone of configurations at which agonist and antagonist muscles are active simultaneously. Thereby, the opposing muscle groups produce balanced torques such that the configuration R remains unchanged. In this sense this command is functionally independent of the r command with the following reservation: since the coactivation zone is created around the R configuration and thus travels with it (Levin & Dimov 1997). In the presence of co-activation, the referent configuration no longer represents a purely threshold configuration but may be more generally regarded as the configuration at which the EMG activity of the muscles show minimal activity with the depth of the minimum defined by the magnitude of the co-activation command.

A testable consequence of the referent configuration hypothesis is now presented. In the presence of inertial and reactive forces, changes in R specified by the nervous system cannot be replicated in the actual configuration. However in movements with reversals, the two configurations might approach and match each other although transiently. If such a matching is present, then the global EMG activity of all working muscles should be simultaneously minimal, with the minima's depth being defined by the degree of co-activation present.

This hypothesis has been tested in three different scenarios which are presented here. We will present results from head movements in monkeys, sit to stand movements with reversals and arm-trunk movements with reversals where a global EMG minima was observed at the point of reversal.

5. Head movements in monkeys

Lestienne et al. (2000) presented targets to monkeys seated in a primate chair with the trunk and abdomen restrained by a seat belt. In a block of 20 trials when the monkey faced the centre of an arc (0°) a fruit was displayed to the left (60°). The monkey turned the head and when the movement approached 60° the fruit was quickly moved to the centre of the arc. Following the target the monkey made a 60° movement to the right. In another block the monkey made an 80° head movements, from an initial head position of 20° to the left to and from a target placed at 60° to the right. Three dimensional coordinates of the monkey's head were recorded along with EMG activity of left and right neck muscles in the large superficial, long and sub-occipital areas linking the skull to the shoulder girdle, vertebral column and the first two vertebrae. The results from the experiment are shown in Figure 3.

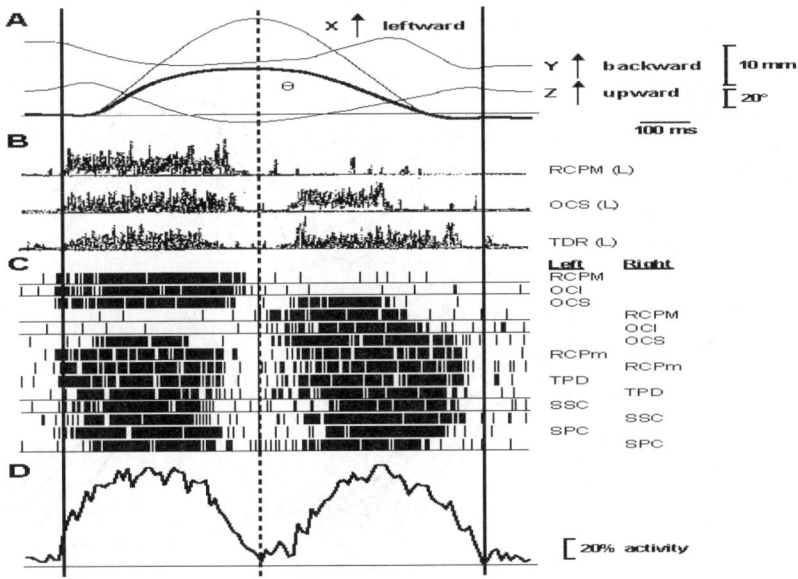

Figure 3. Monkey head movements. Kinematic and EMG patterns of head movements of a monkey during a single movement to and from the left target. B. Typical EMG patterns from three different muscle groups. C. Quantified integral of the EMG activity of seven pairs of neck muscles. D. Ensemble activity of normalised EMG signals of 14 muscles. Reproduced from Lestienne et al (2000) with permission.

An inspection of Figure 3 shows that the EMG activity of all muscles, irrespective of biomechanical function, showed minima at two phases of head movements. One minimum occurred before the onset of the movement, at static positions corresponding to the resting positions at which forces provided by small tonic activity of the neck muscles and by the passive elastic forces of the head-neck system were sufficient to balance the head gravitational torque. More importantly, an additional EMG minimum or a global minimum lasting 70-100 ms was observed during the movement reversal, followed by activity of neck muscles that guided the head-neck system to its initial configuration, as shown in Figure 4. Despite the biomechanical, anatomical and functional diversity of muscles involved, all showed minimum EMG activity at the point of reversal.

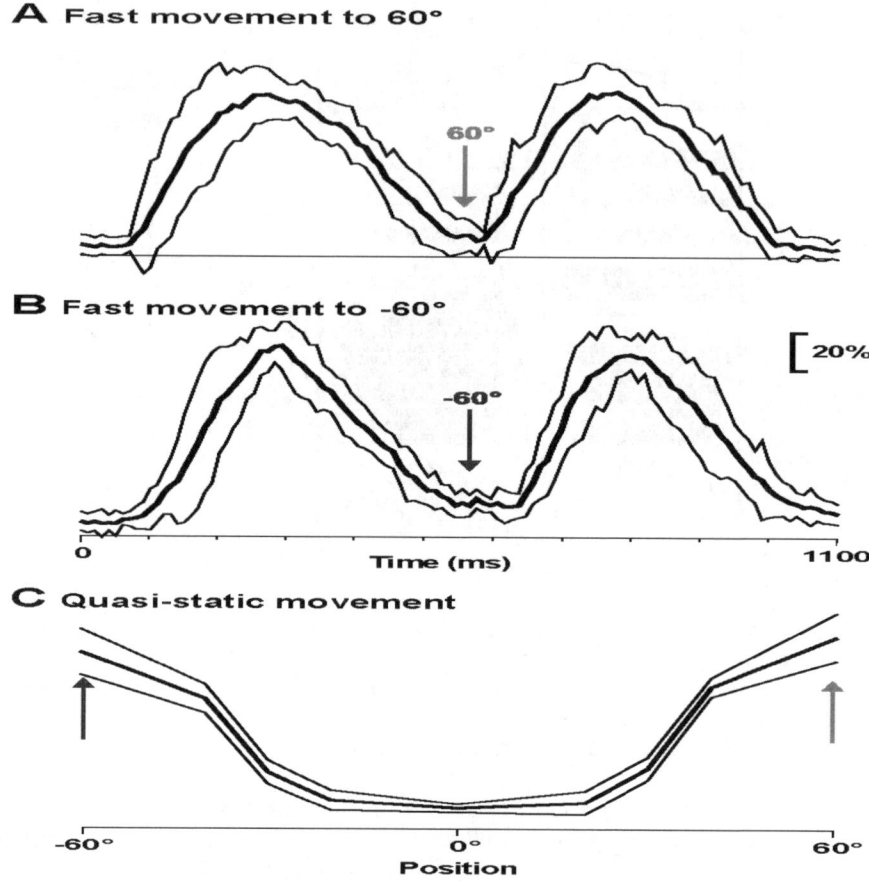

Figure 4. Global EMG minima. Ensemble activity of normalised EMG activity (mean and SD) of 14 muscles during fast movements to the left target (A) and right target (B). During fast movements ensemble activity is minimal at head positions shown by solid arrow in A and open arrow in B, where the activity would be maximal if the appropriate positions were maintained as shown in panel C. Reproduced from Lestienne et al (2000) with permission.

6. Sit-to-stand movements

Feldman et al (1998) investigated a functionally similar movement task with reversals in human subjects. In response to an auditory signal, six healthy subjects sitting on a stool with their arms resting on their knee, rose from the seat to a semi-standing position and returned to sitting. While lifting their body, subjects stretched their arm pointing (without touching) to a plastic disc target placed in front of them at a height of 100cm. The movement was performed smoothly at a comfortable speed, in each of 10 trials (4-5.5s each, 15s between trials). 3-D coordinates of the right shoulder, hip, knee and ankle were recorded in addition to EMG activity in arm, trunk and leg muscles. The results are summarized in Figure 5.

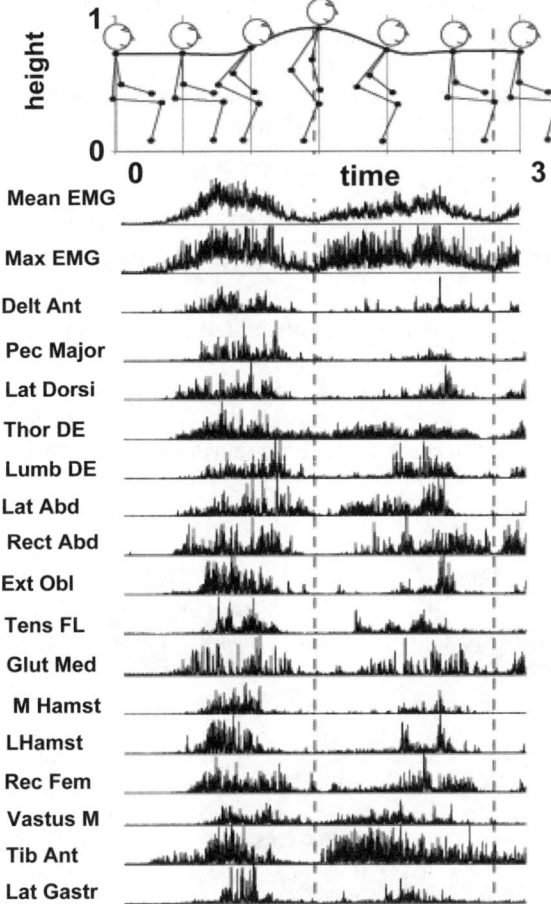

Figure 5. Sit-to-stand movements. An example of normalised EMG signals from arm, trunk and leg muscles and the vertical coordinate of the shoulder during the sit to stand movement with direction reversal. The vertical line shows where the global minimum was observed. The second EMG minimum associated with sitting is shown by the second vertical line. Adapted from Feldman et al (1998) with permission.

Global minima of EMG activity associated with the reversal phase were observed in not only the postural stabilizers and prime movers (trunk and leg muscles) but also in the shoulder muscles, anterior deltoid and pectoralis major. A closer look at Figure 5 shows that, although in some muscles, the global minimum occurred during the brief silent period between sequential bursts, the activity of many other muscles terminated substantially before and resumed comparatively long after the transitional global minimum. The finding of the global minima at the transition phase is thus consistent with the hypothesis that changes in the referent body configuration may underlie sitting to standing movements of the body. We believe that these findings add to the richness of the concept of synergy or manifold as presented by Scholz et al (1999) for a similar task and as described in Schöner (this volume).

7. Trunk assisted pointing movements with reversals

Trunk assisted pointing movements (as illustrated in Figure 6) may be considered the superposition of two synergies; a transport synergy (for arm movement) and compensatory synergy (for changes in shoulder and elbow joint angles). The superposition of the two synergies ensures an invariant trajectory when active trunk flexion is involved in reaching for a target (Pigeon et al, 2000).

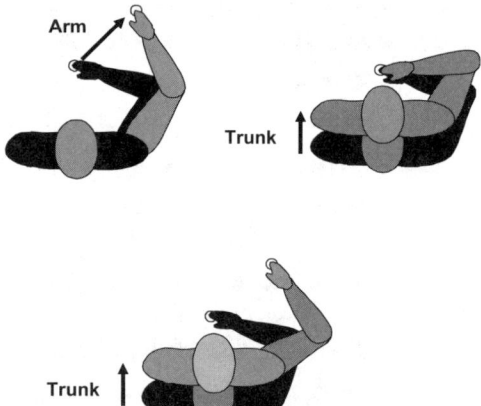

Figure 6. Superposition of synergies. A. Transport synergy for pointing or reaching movement where only the arm moves. B. Compensatory synergy as shown by changes in elbow and shoulder joint angles when bending the trunk forward while maintaining arm posture. C. Superposition of synergies shown in A & B, when the arm moves towards a target ipsilaterally in conjunction with trunk motion where the arm trajectory remains invariant despite additional degrees of freedom as demonstrated by Pigeon et al (2000). Figure adapted from Pigeon et al (2000) with permission.

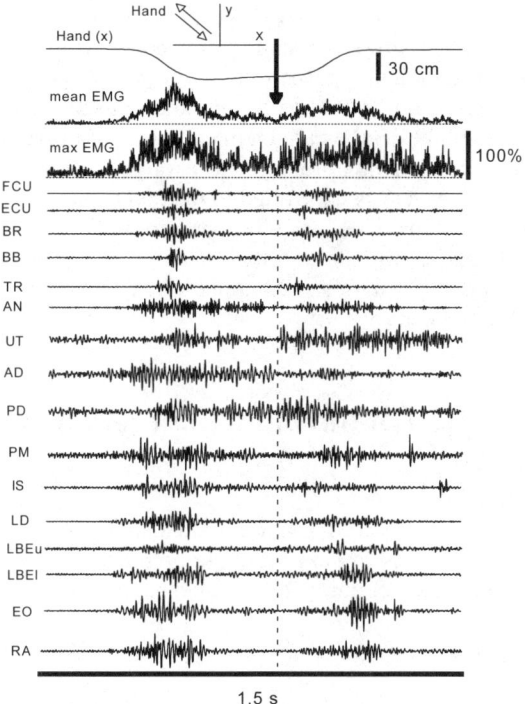

Figure 7. Arm-trunk coordination in reaching. The vertical arrow and dashed line show the occurrence of a global EMG minimum in the activity of 16 muscles of the arm and trunk; diagonal arrow show the direction of the hand movement (x is frontal and y is sagittal direction). EMG of muscles (from top to bottom): flexor and extensor carpi ulnaris, brachioradialis, biceps brachii, triceps,anconeus, anterior and posterior deltoids; pectoralis major, infraspinatus, latisimus dorsi, low back extensors (upper part), low back extensors (lower part), external oblique, rectus abdominus.

Balasubramaniam & Feldman (2001) presented targets in both the ipsilateral and contralateral directions from six seated subjects who had to point to the targets (within reaching distance) with redundant use of the trunk (bending the trunk simultaneously) and reverse the trajectory on reaching the target, akin to the lower panel in Figure 6. The targets were placed within reach in the subject's workspace. 3-D positions of the sternum, shoulder, elbow, wrist and endpoint trajectories along with EMG activity in all major arm and trunk muscles were recorded. As shown in Figure 7, at the point of reversal all the muscles of the arm, trunk synergy showed a global minimum irrespective of their biomechanical function, once again confirming the predictions of the referent configuration hypothesis. One can say that when producing global EMG minima, the neuromuscular system strengthens the role of passive muscle properties and external forces, including gravity, in movement production. For example, during horizontal head movements in monkeys, the EMG minimization at the lateral positions allows the head movement to be reversed by the passive forces accumulated in previously stretched antagonist muscles. In sit-to-stand movements, EMG minimization allows the body to use gravitational forces to assist in returning to sitting. These explanations of the occurrence of global EMG minima in terms of mechanics are not alternative to the R hypothesis: they are an integral part of it. At the same time, taken separately, these explanations do not go far enough to answer the question of how the neuromuscular system is controlled to take advantage of the body and environmental forces and thus mechanically optimise the movement. The R hypothesis directly addresses this issue: the mechanical optimisation is not intentionally programmed but emerges each time when the actual and referent configurations match each other.

8. Testing the solution of the multi-joint redundancy problem

Like the solution of the problem of muscle redundancy, the solution of the problem of DF-redundancy leads to some testable predictions. We describe two predictions that have been tested in pointing movements involving the trunk.

1. Hand trajectories remain invariant despite changes in the number of DFs involved.
As described above, pointing movements might result from control signals influencing neurons that receive afferent inputs related to the position of the arm endpoint in space and projecting either indirectly, via interneurons, or directly to motoneurons of all skeletal muscles. These control signals represent the referent coordinates of the endpoint. The neurones are activated depending on the difference between the actual and the referent coordinates. According to the principle of the minimal interaction, the system is driven, both locally and globally, to a state at which the difference becomes minimal, in the limits determined by intrinsic (neural) and extrinsic (e.g. gravity) constraints. Thus, control levels produce a referent trajectory of the endpoint whereas the actual endpoint trajectory emerges following the laws of mechanics and the principle of the minimal interaction tending to diminish the difference between the actual and the referent trajectory. This does not mean that the system will eventually establish the position coinciding with the referent one. Rather, the system will arrive at a position that is distant from the referent position so that the difference between these positions will provide the activation of neurons and motoneurons that is necessary to generate muscle forces counteracting the gravitational and load torques acting upon the arm in this position. The difference between the actual and the referent positions might be substantial in the cases when the movement is blocked (isometric torque exertion): since the actual position remains the same, the changes in the referent position will directly influence its distance from the former, resulting in isometric EMG and torque generation.

Guided by the minimisation principle, the system has the capacity not only to bring it to a steady (equilibrium) state but also to resist intermittent mechanical perturbations and thus rapidly return the endpoint trajectory to that obtained in the absence of perturbations. Such kind of dynamical stability has been demonstrated by Won and Hogan (1995). The principle of minimal interaction provides system's stability in a more broad sense: up to specific limits (that are exemplified below), the actual movement trajectory may remain the same regardless of changes in the number of DFs involved in the motor task, whether these changes are produced voluntarily, by the subject, or involuntarily, by imposing or releasing some DFs from mechanical constraints.

This prediction has been confirmed in several studies of pointing movements involving the trunk. In cases of movements to targets placed within the arm's reach, the endpoint trajectory remained invariant when subjects intentionally involved the trunk or when the trunk was recruited but its motion was mechanically blocked (Adamovich et al, 2001). In pointing to targets placed beyond the arm's reach, the trunk contributed to the movement extent and therefore mechanically preventing the trunk motion made reaching the target impossible. However, even in this case, the hand trajectory remained invariant until the hand approached the limits of the movement extent available for the arm alone (Rossi et al. 2002). In all cases, in response to the trunk arrest, the trajectory was maintained due to appropriate short-latency (about 50 ms) changes in the arm joint angles as shown in Figure 8.

It is also interesting to note that trunk-assisted reaching involves different segments that move in parallel when central commands related to functional components of the task are generated sequentially, akin to those in speech production. It is normally observed that commands underlying speech production are generated sequentially but the motor responses overlap, resulting in an acoustic effect called co-articulation (Fowler & Saltzman, 1993). This sequential control structure is similar to the problem of "serial order in behaviour" as discussed by Lashley (1951). He noticed that components of different behaviours including locomotion, prehension and language are generated sequentially and such generation cannot result from moment-to-moment responses to a serially ordered environment. Serial order is a manifestation of a control principle that guides the behaviour of the organism, much the same way we propose that movements maybe guided without redundancy.

2. Hand trajectory invariance may be provided in different FRs, depending on task demand.
Another consequence of the solution of DF-redundancy problem is related to the concept of FR. When we discussed the production of pointing movements we implicitly assumed that, since the targets were placed on a table, the referent coordinates of the endpoint were specified by the nervous system in a FR associated with the environment (experimental room) and, respectively, the invariance of the endpoint trajectory was observed in this FR. The invariance was maintained by substantial changes in the arm interjoint coordination in response to

the trunk arrest. However, one can produce pointing to a target on the body (e.g. to the nose) so that the target will move with the trunk. To reach the target, the referent coordinates of the arm endpoint should be shifted in a FR associated with the trunk. One can predict that, the trajectory of pointing to such target will remain the same whether or not the motion of the trunk was prevented. Trajectory invariance, however, will be observed not in a FR associated with the environment

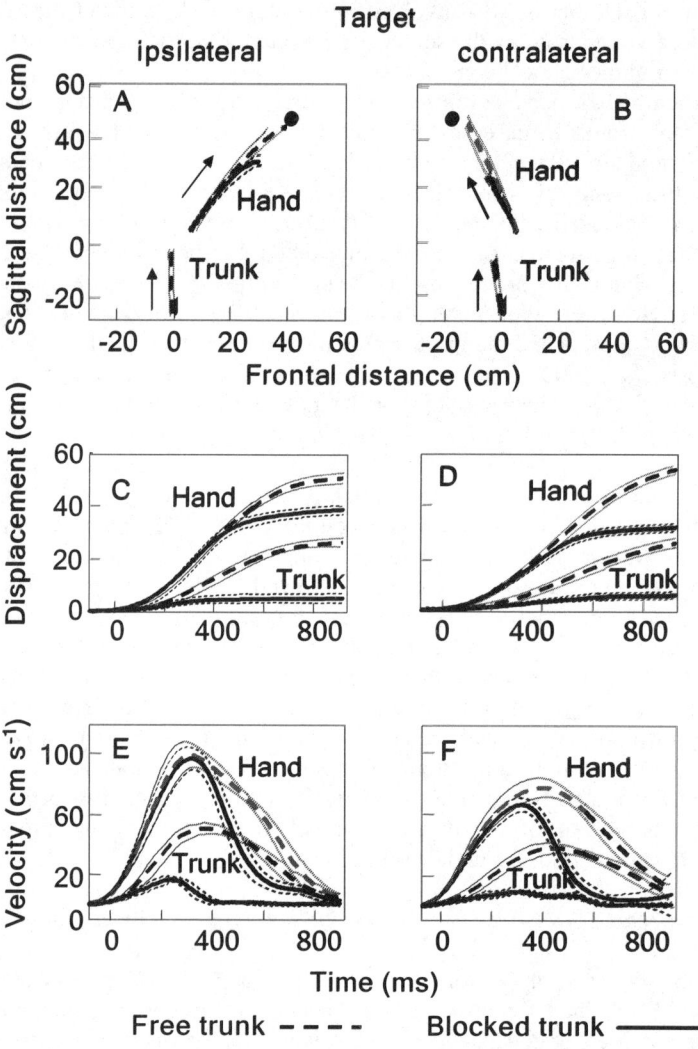

Figure 8. Sequential activation of synergies. Typical kinematic effects of trunk arrest in movements to targets located beyond the reach of the arm. Mean traces with SD for conditions when trunk motion is not obstructed (thick dashed lines) or blocked (thin dashed lines); arrows in A and B show movement directional and filled circles show targets. In free

trunk movements subjects leaned the trunk forward about 23 cm and these displacements were reduced by about 3 cm when trunk movement was mechanically blocked. Despite differences in the trunk motion, the endpoint trajectories (A, B), displacements along them (C, D) and velocity profiles (E, F) initially followed the paths in trials in which the trunk was free to move. Adapted from Rossi et al. (2002) with permission.

but in a FR associated with the trunk. In addition, in contrast to pointing to targets in the environment, no change in the inter-joint coordination will be necessary to keep the trajectory invariant despite the trunk arrest. The ability to produce pointing movements in two different FRs with substantially different behaviours at the joint levels has been confirmed in a recent study (Ghafouri et al. 2002). This study also shows that the transformation of the motor performance from one FR to another does not require learning, suggesting that FRs are pre-existing structures and it remains for the nervous system to choose an appropriate ("leading") FR to control movement.

9. Synergies (primitives, controlled and uncontrolled manifolds) as emergent characteristics of motor behavior

In the previous section describing the control of pointing movements, we emphasised the capacity of control levels to select task-specific FRs with the origin representing the referent coordinates of the endpoint. When one of the FRs is chosen as leading, it subordinates the FR comprised of all-possible body configurations by controlling its origin – the referent body configuration. The configurational FR, in turn, subordinates local FRs associated with single muscles and eventually motor units. Our solution of the DF-redundancy problem implies that control levels may guide the motor performance by shifting the origin of the leading FR whereas appropriate referent body configurations and eventually individual values of muscle thresholds emerge automatically. Moreover, while control levels may reproduce the same referent trajectories at the leading level, the emerging referent body configurations and, as a consequence, actual body configurations may change with the task conditions (like was the case in response to trunk arrests). By analysing these movements in terms of kinematics one can find appropriate functional subdivisions of DFs (synergies, primitives or manifolds), as for example was found for pointing movements involving the trunk or pistol shooting (Scholz et al, 1999, 2000, 2001). One can consider these subdivisions as a property emerging following the tendency of the system to minimise the activity and interactions in the system including the environment.

Using the λ model, we have shown that movements may be guided without redundancy in a non-computational, dynamical manner. Previous attempts to solve the redundancy problems such as force control models have been unsuccessful as they have always been formulated in terms of output variables (EMG, forces, and kinematics) with which the control functions of the nervous system are not directly associated. We believe that the referent configuration which is grounded in physiology may be very important in guiding multiple muscle systems without redundancy in the production of movements. Control in such systems is achieved by taking advantage of a position dependent attractor, which dictates the organization of moving parts into synergies (or manifolds as the case may be) and makes coordination possible. Coordination dynamics is the study of the formation and evolution of such an attractor (Kelso, 1995). The stable states of these systems are controlled by system parameters and not forces.

References

Asatryan DG, Feldman AG (1965) Functional tuning of the nervous system with control of movement or maintenance of a steady posture: 1. Mechanographic analysis of the work of a limb on execution of a postural task. Biophysics 10, 925-935

Adamovich SA, Archambault PS, Ghafouri M, Levin MF, Poizner H, Feldman AG (2001) Hand trajectory invariance in reaching movements involving the trunk. Exp Brain Res 138, 288-303

Balasubramaniam R, Feldman AG (2001) Frames of reference in reaching movements with reversals. In: Proceedings of the XI International conference on Perception and Action, Storrs, CT

Balasubramaniam R, Feldman AG (2002) Some Robotic imitations of biological movement systems might be counterproductive. Behav Brain Sci 24, 1050-1051

Balasubramaniam R, Wing AM (2002) The dynamics of standing balance. Trends Cogn Sci 6, 531-536

Bernstein N (1967) The coordination and regulation of movements. Pergamon Press, Oxford

Bhushan N, Shadmehr R (1999) Computational nature of human adaptive control during the learning of reaching movements in force fields. Biol Cybern 81, 39-60

Feldman AG, Levin MF (1995) The origin and use of positional frames of reference in motor control. Behav Brain Sci 18, 723-806

Feldman AG, Orlovsky GN (1972) The influence of different descending systems on the tonic stretch reflex in the cat. Exp Neurol 37, 481-494

Feldman AG, Levin MF, Mitniski AM, Archambault P (1998) Multi-muscle control in human movements. J Electromyogr Kines 8, 383-390

Fowler C, Saltzman EL (1993) Coordination and coarticulation in speech production. Lang Speech 36, 171-195

Gelfand IM, Tsetlin ML (1971) On mathematical modelling of mechanisms of central nervous system. In: Gelfand IM, Gurfinkel VS, Fomin SV, Tsetlin ML (eds.) Models of structural-functional organization of certain biological systems. MIT Press, Cambridge, MA

Ghafouri M, Archambault P, Adamovich SV, Feldman AG, (2002) Pointing movements may be produced in different frames of reference depending on task demands. Brain Res 929, 117-128

Glansdorff P, Prigogine I, (1971) Thermodynamic Theory of Structure, Stability and Fluctuations. Wiley, London

Hollerbach JM (1972) Computers, brains and the control of movement. Trends Neurosci 6, 189-192

Kawato M (1999) Internal models of motor control and trajectory planning. Curr Opin Neurobiol 9, 718-727

Kelso JAS (1995) Dynamic Patterns. Cambridge, MIT Press

Lashley KS (1951) The problem of serial order in behaviour. In: Jefress LA, (ed.) Cerebral mechanisms in behaviour. Wiley, New York

Lestienne FG, Thullier F, Archambault P, Levin MF, Feldman AG (2000) Multi-muscle control of head movements in monkeys: The referent configuration hypothesis. Neurosci Lett 283, 65-68

Levin MF, Lamarre Y, Feldman AG (1995) Control variables and proprioceptive feedback in fast single-joint movement. Can J Physiol Pharm 73, 316-330

Levin MF, Dumov M (1997) Spatial zones for muscle co-activation and the control of postural stability. Brain Res 757, 43-59

Levin MF, Selles RW, Verheul MHG, Meijer OG (2000) Deficits in coordination of agonist and antagonist muscles in stroke patients: Implications for motor control. Brain Res 853, 352-369

Matthews PBC (1959) The dependence of tension upon extension in the stretch reflex of the soleus muscle in the decerebrate cat. J Physiol 147, 52-546

Ostry DJ, Feldman AG (2003) A critical evaluation of force control hypothesis in motor control. Experimental Brain Res, in press

Pigeon P, Yahia LH, Mitniski AB, Feldman AG (2000) Superposition of independent units of coordination during pointing movements is preserved in the absence of visual feedback. Exp Brain Res 131, 336-349

Rossi E, Mitniski AM, Feldman AG (2002) Sequential control signals determine arm and trunk contributions to hand transport during reaching in humans. J Physiol-London 538, 659-671

Scholz JP, Schöner G (1999) The uncontrolled manifold concept: Identifying control variables for functional tasks. Exp Brain Res 26, 289-306

Scholz JP, Reisman D, Schöner G (2001). Effects of Varying Task Constraints on Solutions to Joint Control in Sit-to-Stand. Exp Brain Res 141, 485-500

Scholz JP, Schöner G, Latash ML (2000) Identifying the control structure of multijoint coordination during pistol shooting. Exp Brain Res 135, 382-404

Von Holst E, Mittelstaedt H (1950/1973) Daz reafferezprincip. Wechselwirkungenzwischen Zentralnerven-system und Peripherie, Naturwiss, 37 467-476. The reafference principle. In: Martin R (translator) The Behavioral Physiology of Animals and Man. The collected papers of Erich von Holst. University of Miami Press, Coral Gables, Florida

Wolpert DM, Ghahramani Z (2000) Computational principles of movement neuroscience. Nat Neurosci 3, 1212-1217

Wolpert DM, Ghahramani Z, Flanagan JR (2001) Perspectives and problems in motor learning. Trends Cogn Sci 5, 487-494

Wolpert DM, Kawato M (1998) Multiple paired forward and inverse models for motor control. Neural Networks 11, 1317-1329

Won J, Hogan N (1995) Spatial properties of human reaching movements. Exp Brain Res 107, 125-136

A Perceptual-Cognitive Approach to Bimanual Coordination

Franz Mechsner

Max Planck Institute for Psychological Research, Munich, Germany

1. Introduction

It is a matter of much debate whether voluntary movement performance and learning takes place in a perceptual-cognitive medium only or relies on an additional level of coordinative processes in the motor system. Here, I will argue in favor of the "perception and cognition only" working hypothesis, which says that there is no level or stage in human motor control where coherent muscular activity patterns are organized as such. Instead I propose that human movements are planned, executed, and stored in memory by addressing their anticipated perceptual consequences. Many factors including physical, biomechanical and neuro-muscular ones influence these consequences. However, the criteria ruling the coordinative action are those of the perceptual-cognitive system, in the first place. I present some experiments on bimanual interference whose outcome suggest that these phenomena are perceptual-conceptual in nature.

2. Self-organization and control in biology and psychology

Taking Charles Darwin's (1859) notion seriously that organisms, and thus the abundantly complex universe of life, have evolved on the simple basis of mutation and natural selection means appreciating that living matter and the processes of life are self-organized. Darwin could imagine that God might have created one or several simple primeval organisms, as a starting point. Eigen (1971; Eigen & Schuster, 1979) radicalized his notion by mathematically showing that life-like, self-sustaining autocatalytic dynamic systems far from thermodynamic equilibrium (the so-called hypercycle) could emerge, in principle, from "dead" matter, i.e., suitable mixtures of molecules. In consequence, it seems most plausible to conceive living nature as a self-organized dynamic material pattern. If so, it is trivial to recognize that all biological objects as well as biological events, including mental phenomena and overt behavior, result from self-organizational processes.

Nevertheless, for heuristic reasons, it makes sense to identify levels of "control" in biological systems as distinguished from levels of "self-organization". The notion of control levels appreciates that the system can behave in a self-sustaining and adaptive way (sometimes called "teleonomic", which means quasi-goal-directed, e.g., Mayr, 1982). However, what is to be considered a level of control under one point of view is to be considered a level of self-organization under another point of view (see Kelso, 1995). To give an example, for considering molecular dynamics in a liver cell, DNA may rightly be regarded as a control structure whose existence is taken for granted independent of the fact that its information has emerged by way of self-organizational processes, in the course of evolution. Obviously, choosing the appropriate point of view is essential in considering biological phenomena with regard to self-organization and control.

3. Perceptual-cognitive control in human action

It seems obvious that in human behavior there are specific levels of control, as for instance, in connection with "goals" and "intentions", or, more generally speaking, with mental processes and states. As understood here, these control levels provide representational reference structures leading the system into the corresponding states. Seen as such, these reference structures can be considered anticipatory. This is independent of whether one conceives these representations as providing internal control parameters, so-to-speak (Keijzer, in press; Kelso, 1995), which guide self-organized executional processes, or whether one conceives these representations as reference structures in behavioral servo-loops which guide adaptively tuned executive signals in a way that the difference between anticipated, or desired, and actual state is diminished (Powers, 1973). In any case, it is of much theoretical interest whether and to which extent human behavioral patterns emerge in service of and adapted to such representational reference structures, or whether and to which extent they mirror processes on the executional level, the unbound influence of biomechanical and physical constraints and the like, apart from cognitive-perceptual control. It seems not implausible to assume, that the cognitive system continuously, and often very quickly learns, about these constraints and takes them into consideration. In consequence, the coordinative action is not governed by "blind" physical, biomechanical, and neuro-muskulo-skeletal factors but by perceptual-cognitive rules, i.e., by perceptual and conceptual grouping principles, strategies minimizing working memory load, and so on. Of course, such perceptual-conceptual control is never complete, but it may be much more thorough than often has been assumed, especially in the case of comparably simple movements.

4. The symmetry tendency in bimanual finger oscillation

4.1. The classical bimanual finger oscillation paradigm

Spontaneous coordination tendencies such as the so-called symmetry tendency in bimanual movements are of special interest in this regard. One may hope that understanding their origin and mechanism will lead us a step further towards understanding the basic mechanisms of movement control, in particular with regard to the functional locus of movement organization, control, and learning. It is not in the least clear whether the symmetry tendency and related phenomena are perceptual-cognitive or motoric in nature. (Here, I define "motoric" processes as processes in executional neuronal structures such as cross-talk between efferent neuronal command streams, as distinguished from perceptual-cognitive processes such as planning operations. Though not well-defined, this rough distinction may do for the present purpose.)

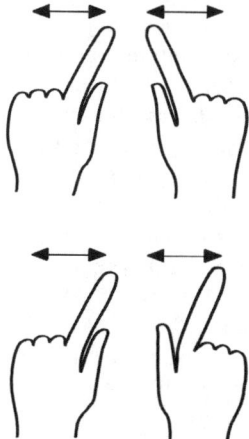

Figure 1. Instructed synchronous finger oscillation patterns in Experiment 1. Symmetrical movements (A), and parallel movements (B).

In this paper I will focus on the well-known bimanual finger oscillation model as introduced by Cohen (1971) and Kelso (1981, 1984). I consider it with regard to the above mentioned key-question: Are movements organized in a motoric or rather in a perceptual-cognitive functional domain? The symmetry tendency in this experimental model can easily be observed: A person places his or her hands in a sagittal direction, in mirror-symmetry to the sagittal body midline, and oscillates the index fingers in the transverse plane (Figure 1).

A mirror-symmetrical oscillation pattern is stable up to the highest possible oscillation frequencies. In striking contrast, an instructed parallel oscillation pattern is stable only at slow and moderate speeds. With increasing oscillation frequencies, the parallel pattern disintegrates, and the fingers often spontaneously switch in a symmetrical movement pattern. The startling stability phenomena observed in this and similar bimanual coordination paradigms have stimulated, above all, the development of the dynamic systems approach to movement understanding (Haken, Kelso & Bunz, 1985).

4.2. The traditional homologous muscle approach is misleading

What is the functional origin of the symmetry tendency? Is it motoric-efferent in origin or rather perceptual-cognitive? Beginning with the papers by Cohen (1971) and Kelso (1984) it has often been emphasized that symmetric movements of homologous limbs involve synchronous contractions of homologous muscles. The wide-spread habitual reference to muscular synchronization patterns in discussions of bimanual coordination dynamics is understandable in the light of the traditional propensity towards identifying the problem of motor control with the problem of bringing about the appropriate spatial-temporal muscular activation pattern.

However, this habit is rather misleading, in several respects. First, it often prematurely implies or suggests an interpretation of the symmetry tendency, which excludes possible explanations in a non-anatomical spatial frame of reference. Second, activation of any muscle is confounded with movement characteristics, which might be the focus of organization, rather than muscular activation itself. Third, it suggests that there is a fundamental homologous muscle "principle" at work, which is of outstanding and particular relevance here, maybe mirroring canonical bilateral cross-talk in efferent motor pathways or the like which may suffice as an explanation. However, it is not in the least clear whether such a principle exists.

In the following, I will first show for the case of one paradigm, namely periodic bimanual index finger adduction and abduction, that the symmetry tendency here is certainly not towards co-activation of homologous muscles. The experimental logic on which I will rely is simple: If it is possible to show that the symmetry tendency is not a tendency toward co-activation of homologous muscles, the idea that it is due to a dominant co-activation tendency of homologous neuronal motor pathways could firmly be refuted. Instead, the hypothesis would become plausible that the symmetry tendency in this paradigm is perceptual-cognitive in nature. (It is very important to note that only evidence against a homologous-muscle approach would be telling with regard to our question of a motoric-efferent versus perceptual cognitive origin of the symmetry tendency. If the most stable mode would involve a co-activation of homologous muscles, this would leave open that question, as a co-activation tendency of homologous muscles might arise on a motoric-executional level as well as on a perceptual-cognitive parameterization level.) After reporting this experiment I will go to another paradigm, namely bimanual index finger flexion and extension. Here, in striking contrast to the adduction and abduction paradigm, homologous muscles are co-active in the most stable oscillation mode. However, I will argue and provide some evidence that also in this case perceptual principles apart from muscular homology are governing coordination.

4.3. Coordination tendencies in periodic index finger adduction and abduction

4.3.1. The symmetry bias in bimanual index finger adduction and abduction is spatially defined

The following experiment, which I will refer to as Experiment 1a here, has been reported in Mechsner, Kerzel, Knoblich, and Prinz (2001, Experiment 1). I will only sketch the main findings.

Participants (n=8) performed adductive-abductive bimanual index finger oscillations in the transverse plane. The hands, placed parallel, in mirror-symmetry to the sagittal midline, were individually positioned either palm down or palm up. Two of the four possible palm combinations are "congruous", i.e. with both palms up or both palms down, and two of them are "incongruous", i.e. with one palm up and the other palm down (see Figure 2).

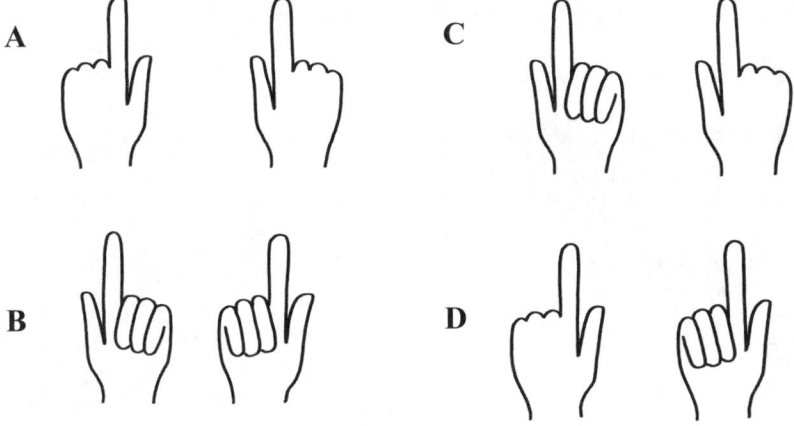

Figure 2. Instructed palm positions in Experiment 1. Congruous positions with both palms up (A) or both palms down (B). Incongruous positions with one palm up and the other palm down (C, D).

There were two different movement instructions (see Figure 1). A "symmetrical" movement instruction requires the participant to oscillate his or her fingers in a mirror-symmetrical fashion with regard to the sagittal body midline (0° relative phase). A "parallel" movement instruction requires the participant to oscillate the fingers in a parallel fashion, i.e., the movement of one fingertip towards the midline goes together with a movement of the other fingertip away from the midline, and vice versa (180° relative phase). A metronome pulse paced the oscillation frequency from 1.4 Hz up to 3.6 Hz in a trial of 24 s duration. If participants felt a tendency to slip out of the instructed pattern they should not resist too much but adopt the more comfortable pattern instead. All participants performed 32 trials, in a 2 palm position (congruous, incongruous) x 2 instruction (symmetrical, parallel) design, in a randomized order.

The crucial condition is defined by the incongruous hand settings. If there is a dominant tendency toward co-activation of homologous muscles, the parallel pattern will be more stable than the symmetrical pattern. In contrast, if there is a dominant tendency toward spatial, i.e. perceptual, mirror-symmetry with regard to the sagittal midline, the symmetrical pattern will be more stable than the parallel pattern.

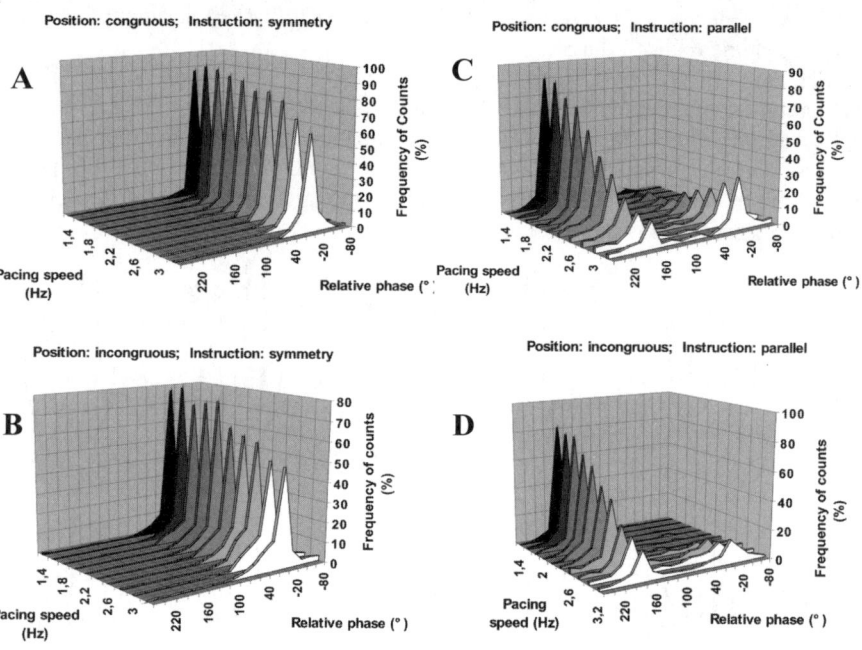

Figure 3. Histogram plots of the proportion of relative phase counts, distributed across 18 relative phase intervals, of 20° each, as a function of metronome frequencies, as revealed in Experiment 1.

The results were clear-cut. Instructed symmetrical movements were quite successfully performed up to the highest movement frequencies, independent of palm position. In striking contrast, parallel movements were correctly performed only at low metronome speed. At increased speed, the parallel pattern dissolved. Moreover, transitions in the symmetrical movement mode were obvious. Figure 3 demonstrates this by displaying histograms of the relative phase of the fingertips (for details of the procedure see Scholz & Kelso, 1989; Lee et al., 1996). For the interpretation of the relative phase axis, it is only important to know here that 0° relative phase means perfect mirror-symmetry, whereas 180° relative phase means perfect parallelity. Relative phase was calculated upon every right reversal of the left finger. The pacing speed axis displays the metronome speed in a 2 s interval. This axis may implicitly be read as a time axis with the metronome pace speeding up over a trial. The count axis displays the percentage of counts of relative phase in every 20° relative phase interval.

Statistical analyses confirm these results (Mechsner et al., 2001). It has to be noted, that virtually the same results are revealed if the participants' view of the hands is occluded, thus they have to rely on kinesthetic proprioception only, with regard to perception. (I refer to this experiment as Experiment 1b.) As kinesthetic proprioception is never "switched off" in normal life, one can firmly conclude that the symmetry tendency in periodic bimanual finger adduction and abduction is organized as a perceptual event, under vision as well as under occluded vision. A tendency towards co-activation of homologous motor pathways can not be responsible for the observed symmetry tendency. The most flexible tuning of muscular activity in service of the symmetry tendency strongly suggests that only the perceptual goal is of importance here and the focus of control, but not the exact pattern of motor commands.

These are challenging results as they do not simply add to the already existing body of evidence that perceptual constraints can play a significant role in influencing or even bringing about spontaneous bimanual coordination phenomena. For such evidence, it has been known before, as for instance, that there is a tendency between two people looking at each other to synchronize oscillatory movements of the limbs as well as of hand-held pendulums. Even switches from parallel to symmetrical movement patterns can be observed in interpersonal coordination (Schmidt, Carello & Turvey, 1990; Schmidt & O'Brian, 1997; Schmidt, Bienvenu, Fitzpatrick & Amazeen, 1998). Similar phase transitions also occur in rhythmic coordination of one single limb with an oscillating visual cue (Wimmers, Beek & van Wieringen, 1992). Swinnen and colleagues (Swinnen, Jardin, Meulenbroek, Dounskaia & Hofkens-Van den Brandt, 1997; Swinnen et al., 1998) demonstrated that movements of the hands which are parallel in extrinsic space have a certain stability advantage relative to other movement patterns in extrinsic space. Kelso, Buchanan and Wallace (1991) demonstrated that the preferred and most stable intralimb coordination pattern involving wrist and elbow movements is independent of the supine or prone position of the forearm and thus of the muscles and motor commands involved. (This list of evidence for a role of perceptual-cognitive constraints in spontaneous movement coordination is by far not complete and may well be lengthened.)

It must be noted, however, that the experiments reported so far do not directly address the most interesting phenomenon in this area, namely the intrapersonal synchronization and symmetry tendency of homologous limbs. Here, it has been argued that motoric constraints might be dominant, due to a co-activation tendency of homologous motor pathways (e.g., Cattaert, Semjen & Summers, 1999; Swinnen, 2002). However, the results of our Experiment 1 show, in a paradigmatic case, where a canonical co-activation tendency of homologous motor pathways should become obvious, that such a tendency plays actually no, or at least no significant, role.

The notion that constraints in the motor system bring about the symmetry tendency has been based on the assumption that there is a general tendency towards co-activation of homologous muscles. In case it turns out that there is, actually, no general tendency towards co-activation of homologous muscles, such theoretical concepts might also seriously be doubted in the case of other paradigms. Instead, one may plausibly hypothesize that spontaneous coordination tendencies of this kind are not motoric but generally perceptual-cognitive in nature. If this hypothesis turned out to be true, coordination between persons and objects, persons and persons, as well as between limbs of one person can be understood in terms of one unifying basic principle, namely that they are of a perceptual-cognitive origin, as a rule.

It remains to be shown, of course, first, that a perceptual-cognitive explanation of spontaneous coordination might also be plausible in other paradigms, and second, which particular perceptual-cognitive factors are responsible for the respective coordination tendencies. As there are a huge many perceptual and conceptual grouping principles, spontaneous coordination phenomena might well mirror this manifold and not be well-labeled by the word "symmetry tendency". To note, symmetry is confounded with many possible perceptual factors, which might be effective here, varying from case to case.

As a first guess, one might assume that movements which are easy to perceive are easy to control as well and argue, in connection, that mirror-symmetry "as such" was such a salient perceptual grouping principle (see Wagemans, 1997) that reference to this salience would suffice as a first explanation. However, this is not so clear. Stating that the symmetry tendency might be perceptual-cognitive in nature, as a rule, does not in the least imply that it is due to one single universal perceptual factor, in all paradigms. For instance, it might well be that the symmetry tendency is not a basic feature of the system, or a first principle, so-to-speak, but the default result of a more general tendency to specify represented movement parameters in the same uniform way for both hands. In the case that all movement parameters are uniformly specified in this way, mirror-symmetry results. In the case one or more parameters are not uniformly specified for both hands, this may lead to spatially "symmetrical" as well as "asymmetrical" movement patterns, dependent of the interplay of the particular movement parameters and goal features which are of relevance in the particular case. However, apart from this impressive surface phenomenon, it might be theoretically more important to realize which parameters are uniformly, and which are differently specified in the particular situation. Thus "symmetry" and "asymmetry" are well-defined only with regard to the individual movement parameters and goal features, but ill-defined as an overall characteristic of the movement.

4.3.2. An unexpected reference frame for the symmetry bias

One may ask, for instance, what would happen in the bimanual finger oscillation task if the hands were not placed in parallel, equidistant to the sagittal midline, but in some other absolute and relative position. In the next experiment (Experiment 2), which is reported in detail in Mechsner (submitted a), participants (n=10) placed their forearms in T-form. This means, one arm was put in a sagittal direction and the other arm in a transversal direction, as sketched in Figure 4. Participants oscillated their index fingers like in the previous experiment, following an increasing metronome pace. The hands were again individually put palm up or palm down, in all congruous and incongruous combinations, like in Experiment 1. There were two instructed movement patterns. The "mid-down" mode requires that the finger of the sagittal hand goes towards the sagittal midline while the finger of the transversal hand goes towards the body. The "mid-up" mode requires that the finger of the sagittal hand goes towards the sagittal midline when the finger of the transversal hand goes away from the body.

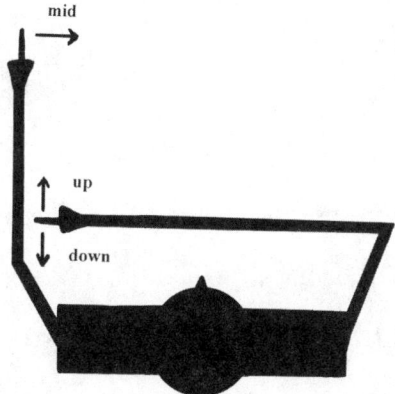

Figure 4. Instructed arm position in Experiment 2. Here, the left arm sagittal, right arm transversal position is displayed. Flipping this picture around the sagittal midline results in the alternative, left arm transversal, right arm sagittal position.

The question of interest was to see whether one of these coordination modes is more stable than the other, and, in particular, whether there is a distinct transition tendency out of one movement pattern to the other. Note, no oscillation mode can be considered mirror-symmetric, thus no symmetry tendency may reveal here. Instead, one might expect that both patterns behave in the same way, be it that they are well maintained up to the highest oscillation frequencies, be it that they will dissolve, or whatever other behavior may be observed. It could also be that a possible tendency towards synchronous performance of anatomically compatible movements would reveal, as such a tendency, if there is one, can not be oppressed by a dominant symmetry tendency here. In any case, it is not in the least obvious what kind of stability behavior should be expected.

Surprisingly, the mid-down mode turned out to be much more stable than the mid-up mode, independent of whether the palm position was congruous or incongruous. Figure 5 shows this by displaying a series of histogram plots analogous to Figure 3. Here, for a definition, 0° relative phase indicates the mid-down mode whereas 180° relative phase indicates the mid-up mode. An instructed mid-down pattern could be quite correctly performed up to the highest speed. In contrast, an instructed mid-up pattern is fairly unstable, from the beginning. At the highest metronome frequencies, it is almost completely dissolved. Even at low metronome frequencies a strong tendency to adopt the mid-down mode instead is obvious. These results have been confirmed by statistical analysis (Mechsner, submitted a).

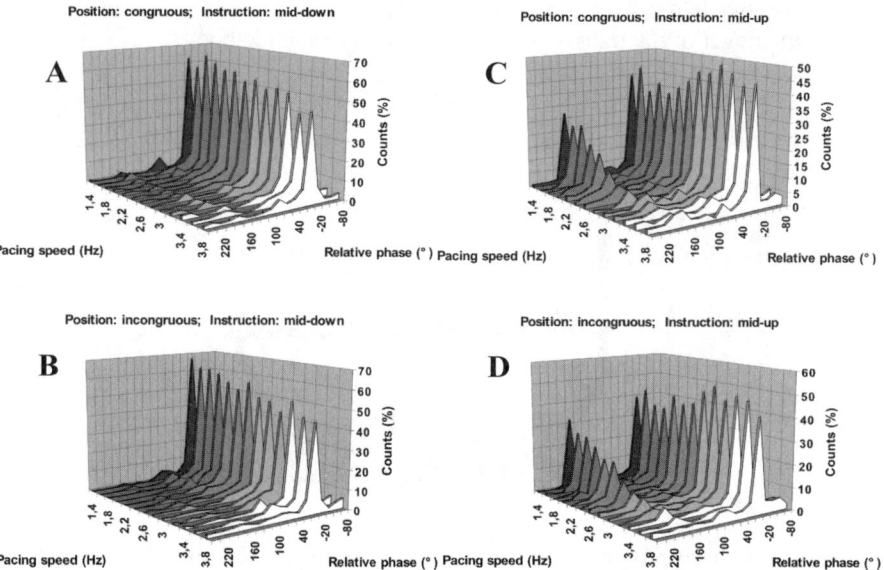

Figure 5. Histogram plots of the proportion of relative phase counts, distributed across 18 twenty degree relative phase intervals as a function of metronome frequencies, as revealed in Experiment 2.

As the revealed mid-down advantage and preference is independent of palm position, it is certainly not defined in muscular coordinates and thus most likely not motoric-efferent in nature. In consequence, I hypothesize that it is perceptually defined. If so, one has to ask how this preferred pattern might be characterized, in perceptual terms.

The preference for a "mid-down" pattern of coordination is reminiscent of the so-called orthogonal stimulus-response (S-R) compatibility which was demonstrated in experiments on proper spatial compatibility (Weeks, Proctor & Beyak, 1995), as well as on the Simon effect (Hommel & Lippa, 1995). S-R compatibility means that there is a reaction time (RT) advantage if the spatial position of the response corresponds to the spatial position of the imperative stimulus, as for instance, if a right-hand stimulus has to be answered by a right response, as compared to a left-hand response. This advantage holds if the spatial position is a relevant stimulus feature ("proper spatial compatibility", Fitts & Seeger, 1953), as well as if it is irrelevant ("Simon effect", see Simon & Rudell, 1967).

Interestingly, an "orthogonal" compatibility effect of the following kind can be found as well: If, for instance, the response has to be made with the right arm placed transversally pointing to the left, there is an RT-advantage for an "up" answer if the imperative stimulus is presented to the right, and an RT-advantage for a "down" answer if the imperative stimulus is presented to the left (Hommel & Lippa, 1995; Weeks & Proctor, 1990). A possible and plausible interpretation is that there is a coordinate system defining "left" and "right" with regard to the respective arm, independent of the actual arm position (Lippa 1996).

When applying analogous considerations to our problem of how to explain the revealed perceptual coordination tendency towards "mid-down" oscillations, one realizes that, in this case, a finger movement "to the right" in one hand is synchronous to a finger movement "to the left" in the other hand. If the hands were placed in parallel, this would mean symmetry. I propose, for a start, to regard the mid-down mode as "orthogonally transformed symmetric" and the mid-up mode as "orthogonally transformed parallel". One may therefore speak of an "orthogonally transformed symmetry tendency" here. It remains a matter of further investigation to evaluate whether the proposed scheme is only phenomenologically justified, or whether there is some deeper justification in the sense that the mid-down versus mid-up and symmetric versus parallel patterns of preference share some underlying common dynamics and mechanisms, or are even, basically, the same.

4.4. Coordination tendencies in periodic finger flexion and extension

4.4.1. The particular stability characteristics in bimanual index finger flexion and extension

In the following I will consider a paradigm, which at first sight seems much similar to the finger adduction and abduction paradigm explored above, namely finger flexion and extension. However, in the case of bimanual index finger flexion and extension surprisingly different stability characteristics have been revealed. In a study by Riek, Carson & Byblow (1992) participants performed a bimanual index finger oscillation task in the transverse plane with their hands either prone or supine. Movements of the index fingers were restricted to flexion and extension. All participants received symmetrical as well as parallel movement instructions, under all four combinations of right and left hand positions. With increasing frequencies, involuntary transitions from the instructed to the alternative mode occurred. With regard to extrinsic space, transitions from the symmetrical to the parallel oscillation mode, as well as transitions the other way around could be observed. With regard to the anatomical character of the finger movement, however, the transition pattern was consistent and unequivocal: Transitions occurred only when flexion in one hand went together with extension in the other hand, in the instructed movement pattern. Thus, after the transition, flexion was always synchronous with flexion and extension with extension. Riek et al. (1992) take this striking result as evidence that involuntary transitions of the described kind are due to a tendency towards co-activation of homologous muscles.

4.4.2. Evidence for a perceptual origin of the flexion-flexion bias

Clearly, homologous muscles are co-active in the most stable movement mode. However, this does not in the least automatically imply that the coordinative principle here is actually defined in terms of muscular homology. And even assumed, that a tendency towards co-activation of homologous muscles was operative here, this would leave open whether it originated on a motoric-efferent or rather on a perceptual-cognitive level. Recently, Oullier, DeGuzman, Jantzen & Kelso (2002) revealed that two people looking at each other while performing periodic unimanual flexion and extension movements of the index fingers tend to interpersonally synchronize flexion with flexion and extension with extension, independent of whether the individual hands were placed palm up or palm down (participants grasped a horizontal stick from downwards or from upwards). As the observed coordination tendency is between persons, it is clearly perceptual-cognitive and not motoric in nature. These results point to the possibility that the flexion-flexion bias might be of perceptual-cognitive origin also in the intrapersonal case.

4.4.3. Is the flexion-flexion bias towards co-activation of homologous muscles?

What perceptual factors might be of relevance here? In the first place, one may argue that a homologous-muscle approach even between persons is not excluded (Oullier et al., 2002). Taken as a hypothesis concerning the operative principle here this claim would mean that muscular homology might actually be the relevant factor guiding coordination. I see only two possibilities to imagine that. First, participants might perceive the relevant muscle activity of the other person and preferably co-activate their own homologous muscles. This possibility can easily be tested. The significant finger flexors and extensors are located in the forearm. If perception of muscular activity was actually crucial for the interpersonal flexion-flexion coordination bias, this tendency should disappear with the hands visible but the forearms covered. My strong guess would be that this will not take place. As a second possibility to invoke a tendency towards co-activation of homologous muscles even between persons it remains to hypothesize that participants, when perceiving flexion or extension of a finger, automatically imagine the muscular activity bringing it about and tune their own muscular activity in a homologous way. This seems not very probable to me though it is certainly possible and should be tested.

In sum, a homologous-muscle approach to inter-personal coordination seems not utterly reasonable, at closer look. In the following I will evaluate whether a perceptual-cognitive explanation of the flexion-flexion and extension-extension synchronization tendency might be found, without invoking muscular homology.

4.4.4. A flexion-flexion bias between non-homologous fingers

In a series of experiments, I tested coordination stability in a periodic bimanual finger flexion and extension task, in close analogy to Riek et al.'s experiments, but this time with non-homologous fingers oscillating together (Mechsner, submitted b). It turned out that there is a strong flexion-flexion bias not only between homologous, but also between non-homologous fingers. This tendency was revealed between index and middle, as well as between index and little finger, while one finger was active in each hand. Though the muscular modules devoted to flexion of the fingers are not well separated, these results suggest that factors apart from exact muscular homology or homology of motor commands bring about the flexion-flexion bias.

4.4.5. A relative salience approach to the flexion-flexion bias

One may extend, or replace, the homologous-muscle approach by hypothesizing that it is not exact anatomical homology as such, but a more general functional similarity of muscles which brings about the flexion-flexion and extension-extension bias. It might be possible that flexors are preferably co-activated with flexors, and extensors with extensors (see Carson, 2003). As for the inter-personal case, the results of Oullier et al. (2002) do not speak in favor of such an interpretation, as I consider it improbable that perception or imagination of muscular activity is of any relevance in the reported paradigm. And it is not implausible to assume that coordination in the inter- and intra-personal are basically governed by the same principles. But even if co-activation of functionally similar muscles is actually what the system focuses on, this cannot not be considered a sufficient explanation, as such. As shown above, in bimanual finger adduction and abduction no tendency towards co-activation of functionally similar muscles (say, adductors with adductors) was revealed. Thus it is, in any case, necessary to Figure out what is special in the flexion and extension paradigm as compared to the adduction and abduction paradigm (for a proposal different from the one I will give here, see Carson 2003).

In the following, I will explore an idea in this regard which I call the relative salience approach to the flexion-flexion bias. I begin with the well-known observation that, in unimanual periodic index finger flexion and extension together with a metronome beat (one movement cycle per beat), there is a clear tendency to synchronize flexion on-beat rather than extension. Moreover, if on-beat extension is instructed, there is a strong bias to switch in on-beat flexion with increasing metronome frequencies but not vice versa (Carson, 1996). Why is this? What is the crucial difference between flexion and extension, which leads to the observed coordination tendency?

I propose, as a possible start, to introduce the concept of relative salience here. Weeks & Proctor (1990) have suggested this concept in order to interpret certain phenomena in connection with stimulus-response compatibility. It has frankly to be admitted that relative salience has only loosely been defined. Colloquially speaking, it is meant that certain perceptual features of an object or event are more likely to catch attention than others as they are more conspicuous, or more stressed. For instance, within green leaves, a red fruit is more salient than a green fruit. A beat is more salient than the pause following it.

In addition, I assume that if a person produces an event with periodically alternating salient and less salient segments he or she tends to synchronize the more salient parts with the more salient parts of any second event with a similar periodicity. For instance, a tap is more salient than the pause between two taps, in a periodic sequence of taps as well as a beat produced by a metronome is more salient than the pause between two beats. In consequence, a person will tend to synchronize his or her taps with the beat, and not with the pause, according to the assumption here. In a way, the more salient part of an event is often perceived as the "real" event, so-to-speak, i.e., as the more important and significant part of the event. This is mirrored in language, as one speaks of a "sequence of beats" rather than of a "sequence of beats and pauses", or of a "sequence of finger taps" rather than a "sequence of tapping and raising movements".

With regard to finger oscillation I propose, as a working hypothesis, that for some reason which has still to be found out flexion is more salient than extension and therefore flexion tends to be performed on-beat and extension off-beat. (This seems to make some sense in daily life. As for instance, grasping an object, which usually involves finger flexion, tends to be perceived as more significant and stressed than extending the fingers in preparation. Or, for another example, taps are spontaneously performed by way of finger flexion rather than extension movements.) If the relative salience approach is correct, periodic flexion rather than extension should be preferably synchronized with the more salient parts of any second event stream which involves a periodic sequence of more and less salient parts. Now have a person produce two event streams, first, periodic flexion and extension of the right index finger, and second, periodic flexion and extension of the left index finger. As flexion is more salient than extension in both hands flexion will tend to be synchronized with flexion. This interpretation provides a possible explanation of the flexion-flexion bias without relying to muscular homology.

The following experiment (Experiment 3) was designed in order to test the prediction that in periodic finger flexion and extension, flexion is preferably synchronized with the more salient parts of any second sequential event stream. Participants were requested to periodically flex and extend their left index finger in the transverse plane, with the hand either prone or supine. At the same time, the

right hand grasped a wooden arrow, with a red head pointing towards or away from the body (see Figure 6).
In the respective direction of the arrow, a yellow mark was presented, as a "goal". Participants were requested to move the arrow periodically towards the goal by way of their right arm and hand, together with flexion and extension of their left finger. Under the assumption that an arrow movement towards the goal is more salient than its movement away from it, participants are expected to preferably synchronize finger flexion with arrow movements towards rather than away from the goal. This is exactly what happens, independent of hand position and arrow direction. (A detailed report is given in Mechsner, submitted c.) This outcome supports the relative salience approach to bimanual finger flexion and extension.
Results by Kelso et al. (2001; see also Carson, 2003) may be taken as further evidence for a relative salience approach. Participants performed unimanual index finger oscillations either according to a flex-on-the-beat or according to an extend-on-the-beat pattern, along with the accelerating pace of an auditory metronome. Coordination patterns were stabilized if the oscillating finger hit an external stop on-beat but destabilized if the finger hit the stop off-beat. Assumed that hitting a stop enhances the perceptual salience of the corresponding finger movement, this means that coordination patterns are stabilized and destabilized according to the relative salience rule proposed above.

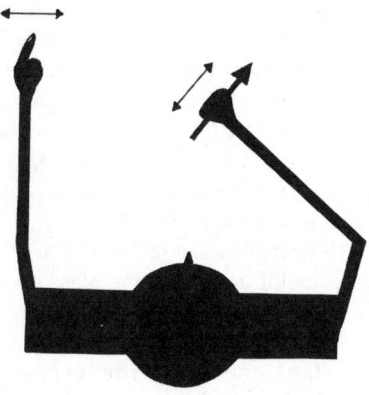

Figure 6. Experimental setting in Experiment 3. The left hand flexes and extends the index finger in the transversal plane whereas the right hand oscillates an arrow pointing outwards (as shown here) or inwards.

5. A perceptual-cognitive approach to action coordination

In conclusion, I propose the hypothesis that voluntary as well as spontaneous coordination is generally perceptual-cognitive rather than motoric-efferent in nature. Obviously, this proposal is rather up for discussion than a proven claim, so far. With regard to spontaneous coordination phenomena such as the symmetry tendency it remains to be seen whether a perceptual-cognitive approach is plausible and can stand the test in an extended range of other coordination paradigms.

Trivially, perception and perceptual anticipation usually have some physical correspondence. Accordingly, a perceptual-cognitive approach to human action does of course not say that factors such as characteristics of the motor system, pre-structured motoric synergies, physical constraints of the body and of manipulated objects are of no importance, quite to the contrary. However, their contribution is not blind, or pure physics, so-to-speak (except it is not predictable), but their influence is integrated in the planning process according to the system's criteria which are perceptually and cognitively defined. As reported in this volume, Carson (2003) introduced manipulations, which changed several properties in the neuro-muskulo-skeletal system. As these manipulations might, in consequence, have changed perceived properties of the investigated movements (such as perceived difficulty and accuracy, relative amplitudes of the ulnar and radial edge of the hand, and so on) his observations do not in the least a priori contradict a perceptual-cognitive approach, as far as I can see.

What is the symmetry tendency in a perceptual-cognitive approach? The keyword in this connection might be dimension overlap. There is a well-established body of research providing ample evidence that feature overlap in perception and action may give rise to massive interference, mainly resulting in facilitation, but sometimes also resulting in inhibition (for a review focusing mainly on perception-action interference see Hommel et al., 2001). Grouping in space and time according to similar or equal features is also possible and common (see Rock, 1984; for a Gestalt approach, see Metzger, 1986).

As has been revealed in the experiments reported above, an overtly much alike, or even virtually identical, looking symmetry bias in different paradigms may actually be due to the influence of very different factors. This has become obvious, in the bimanual index finger oscillation paradigm, by the fact that the simple manipulation of varying hand orientation combinations has led to considerably different coordination tendencies in finger adduction and abduction, as compared to flexion and extension. And even this experimental manipulation was not sufficient in order to arrive at a clear decision, which factors are exactly of importance in these paradigms. Obviously, symmetry is not a simple principle, which is sufficient in itself as an explanation, but may result from the interplay of manifold factors, varying from paradigm to paradigm. In each individual paradigm, the factors of relevance, which are not at all evident a priori, have to be carefully found out and disentangled.

References

Carson RG (1996) Neuromuscular-skeletal constraints upon the dynamics of perception-action coupling. Exp Brain Res 110, 99-110

Carson RG (2003) Governing coordination. Why do muscles matter? This volume

Cattaert D, Semjen A, Summers JJ (1999) Simulating a neural cross-talk model for between-hand interference during bimanual circle drawing. Biol Cybern 81, 343-358

Cohen L (1971) Synchronous bimanual movements performed by homologous and non-homologous muscles. Percept Motor Skill 32, 639-644

Darwin C (1859) On the origin of species by means of natural selection. New edition (1964), Harvard University Press, Cambridge, London

Eigen M (1971) Self-organization of matter and the evolution of biological macromolecules. Die Naturwissenschaften 58, 465-523

Eigen M, Schuster P (1979) The hypercycle: A principle of natural self-organization. Springer, Berlin, Heidelberg

Fitts PM, Seeger CM (1953) S-R compatibility: Spatial characteristics of stimulus and response codes. J Exp Psychol 46, 199-210

Haken H, Kelso JAS, Bunz H (1985) A theoretical model of phase transitions in human hand movements. Biol Cybern 51, 347-356

Hommel B, Lippa Y (1995) S-R compatibility effects due to context-dependent spatial stimulus coding. Psychon B Rev 2 (3), 370-374

Hommel B, Müsseler J, Aschersleben G, Prinz W (2001) The theory of event coding: a framework for perception and action. Behav Brain Sci 24, 849-937

Keijzer F (in press) Self-steered self-organization. In: W Tschacher, JP Dauwalder (eds.), Dynamical systems approaches to embodied cognition. World Scientific, Singapur

Kelso JAS (1981) On the oscillatory basis of movement. B Psychonomic Soc 18, 63

Kelso JAS (1984) Phase transitions and critical behavior in human bimanual coordination. Am J Physiol-Reg I 15, R1000-R1004

Kelso JAS (1995) Dynamic Patterns.The Self-Organization of Brain and Behavior.MIT Press, Cambridge, MA

Kelso JAS, Buchanan JJ, Wallace SA (1991) Order parameters for the neural organization of single, multijoint limb movement patterns. Exp Brain Res 85, 432-444

Kelso JAS, Fink PW, DeLaplain CR, Carson RG (2001) Haptic information stabilizes and destabilizes coordinated movement. P Roy Soc Lond B Bio 268, 1207-1213

Lee TD, Blandin Y, Proteau L (1996) Effects of task instructions and oscillation frequency on bimanual coordination. Psychol Res 59, 100-106

Lippa Y (1996) A referential-coding explanation for compatibility effects of physically orthogonal stimulus and response dimensions. Q J Exp Psychol-A 49, 950-971

Mayr E (1982) The growth of biological thought: Diversity, evolution and inheritance. Belknap Press, Cambridge, MA

Mechsner F (in press) Gestalt factors in human movement coordination. Gestalt Theory

Mechsner F (submitted a) On the symmetry tendency in bimanual finger oscillation

Mechsner F (submitted b) On the flexion-flexion synchronization bias in bimanual finger oscillation

Mechsner F (submitted c) Perceptual coordination tendencies in periodic finger flexion and extension

Mechsner F, Kerzel D, Knoblich G, Prinz W (2001) Perceptual basis of bimanual coordination. Nature 414, 69-73

Metzger W (1986) Gestalt-Psychologie. Kramer, Frankfurt

Oullier O, DeGuzman GC, Jantzen KJ, Kelso JAS (2002) Bimanual coordination between two individuals exhibits homologous muscle constraints. Proceedings of the first conference on Coordination Dynamics, Delray Beach, Florida

Powers WT (1973) Behavior: The control of perception. Aldine, Chicago

Riek S, Carson RG, Byblow WD (1992) Spatial and muscular dependencies in bimanual coordination. J Hum Movement Stud 23, 251-265

Rock I (1984) Perception. Scientific American Books, New York

Schmidt RC, Bienvenu M, Fitzpatrick PA, Amazeen PG (1998) A comparison of Intra- and Interpersonal Interlimb Coordination: Coordination Breakdowns and Coupling Strength. J Exp Psychol Human 24(3), 884-990

Schmidt RC, Carello C, Turvey MT (1990) Phase transitions and critical fluctuations in the visual coordination of rhythmic movements between people. J Exp Psychol Human 16, 227-247

Schmidt RC, O'Brien B (1997) Evaluating the dynamics of unintended interpersonal coordination. Ecol Psychol 9(3), 189-206

Scholz JP, Kelso JAS (1989) A quantitative approach to understanding the formation and change of coordinated movement patterns. J Motor Behav 21, 122-144

Simon JR, Rudell AP (1967) Auditory S-R compatibility: The effect of an irrelevant cue on information processing. J Appl Psychol 51, 300-304

Swinnen SP (2002) Intermanual coordination: From behavioural principles to neural-network interactions. Nat Rev Neurosci 3, 350-361

Swinnen SP, Jardin K, Meulenbroek R, Dounskaia N, Hofkens-Van Den Brandt M (1997) Egocentric and allocentric constraints in the expression of patterns of interlimb coordination. J Cognitive Neurosci 9, 348-377

Swinnen SP, Jardin K, Verschueren S, Meulenbroek R, Franz L, Dounskaia N, Walter CB (1998) Exploring interlimb constraints during bimanual graphic performance: effects of muscle grouping and direction. Behav Brain Res 90, 79-87

Wagemans J (1997) Characteristics and models of human symmetry detection. Trends Cogn Sci 1, 346-352

Weeks DJ, Proctor RW (1990) Salient-features coding in the translation between orthogonal stimulus and response dimensions. J Exp Psychol Gen 119 (4), 355-366

Weeks DJ, Proctor RW, & Beyak B (1995) Stimulus-response compatibility for vertically oriented stimuli and horizontally oriented responses: Evidence for spatial coding. The Q J Exp Psychol-A 48, 367-383

Wimmers RH, Beek PJ, van Wieringen PCW (1992) Phase transitions in rhythmic tracking movements: A case of unilateral coupling. Hum Movement Sci 11, 217-226

Part V: Integration and Segregation in Coordination Dynamics

Complex Neural Dynamics

Olaf Sporns

Department of Psychology, Indiana University, Bloomington, IN 47405

1. Introduction

Brains are large-scale networks consisting of millions of neuronal elements that are interconnected in characteristic patterns. These patterns of anatomical connections are critical for determining which neurons and brain areas can functionally interact. The activation of interconnected neuronal populations gives rise to global dynamical states that are associated with perception and cognition (Bressler, 1995; Frackowiak et al., 1997; Tononi and Edelman, 1998; Mesulam, 1998; McIntosh, 1999; Varela et al., 2001; Jirsa and Kelso, 2003; Ward; 2003). Given the importance of anatomical connections for generating structured neuronal dynamics, we need a deeper understanding of how anatomical connectivity and neuronal dynamics are interrelated. This chapter provides a brief overview of current concepts and models of how structured brain connectivity gives rise to complex neural dynamics. Our discussion will focus on recent results and simulations of the mammalian cerebral cortex.

Each cortical area is connected to a finite and specific set of other cortical areas. Within each area, neurons connect to a small number of neighboring or remote cells, forming local circuits and horizontal cortico-cortical connections. The sum of these connectional relationships, across multiple levels of organization, forms the connection matrix of the cortex, a concise representation of its anatomical connectivity. Compiling data from hundreds of anatomical studies, cortical connection matrices of inter-areal pathways have been derived for several cortical systems, including those of the macaque (Felleman and Van Essen, 1991; Young, 1993) and the cat (Scannell et al., 1999). Several online anatomical databases summarizing these connectional data sets are available (Kötter, 2001). In addition, detailed information on neuronal morphology, local circuitry and horizontal connections is being archived and collated, yielding rich neuroinformatics data sets (Kötter, 2002).

The emerging view is that cortical networks are complex structures with intricate connection patterns ranging in scale from local microcircuits to cortico-cortical and cortico-thalamic pathways extending across the entire brain. What are the principles that underlie these complex patterns? While very few intra-cortical connections and pathways have been studied in detail, some large-scale organizational motifs have been identified. Two major principles of anatomical and functional organization of the mammalian cerebral cortex are segregation and integration. There is overwhelming evidence for anatomical and functional segregation found in virtually all cortical systems, including sensory, motor and prefrontal cortex (reviewed in Tononi et al., 1998). In recent years, the combination of sophisticated anatomical techniques and powerful neuroimaging approaches has lent strong support to the view that the cerebral cortex is subdivided into specialized regions, each containing cell populations that have distinct response and tuning properties and are differentially engaged in various perceptual and cognitive tasks. Anatomically segregated brain areas are interconnected, often reciprocally, by long-range inter-areal pathways, which link remote brain regions into a single large-scale cortical network. These pathways form the structural and anatomical basis for integration of neural activity between distant sites across the cerebral cortex. Segregation and integration, as two of the dominant structural and functional motifs of cortical organization, also underlie the two dominant strategies of neural coding, rate and temporal coding (deCharms and Zador, 2000). Functional segregation is consonant with the tendency of neurons to extract maximal amounts of information from their inputs, a strategy that results in redundancy (and dimensionality) reduction and is best implemented by specialized "feature detectors" encoding stimulus features in terms of firing rate. Functional integration is often associated with a complementary strategy, temporal coding, which operates by creating temporal relationships between distant brain regions, such as temporal correlations or synchrony (Singer and Gray, 1995). Both coding strategies play fundamental roles in neural processing and draw on different structural motifs (segregation and integration) for their neural implementation.

The structure of cortical networks enables the emergence of large-scale patterns of activated and co-activated brain areas, organized into globally coherent, coordinated states (Tononi et al., 1998; Bressler and Kelso, 2001). Patterns of temporal correlations (or, more general, deviations from statistical independence) constitute the system's functional connectivity (Friston, 1994; 1997a), often represented as its covariance or correlation matrix. Numerous studies have shown that multiple distributed cortical areas show consistent and temporally precise patterns of mutual interdependency (correlation or coherence) as specific tasks or cognitive functions are executed. Correlated brain activity and changes in functional connectivity are associated with the perception of Gestalt patterns (Varela et al., 2001), binocular rivalry (Srinivasan et al., 1999), sensorimotor function (Bressler et al., 1993; Roelfsema et al., 1997; Ding et al., 2000), object processing (von Stein et al., 1999), memory encoding and retrieval (McIntosh et al., 1997) and awareness (McIntosh et al., 1999).

2. Complexity and metastability

The pattern of temporal relationships between brain areas (their functional connectivity) represents the ongoing interplay between segregation and integration. These two tendencies may be viewed as "antagonistic", with increased integration disrupting local segregation and *vice versa*. It has been argued (e.g. Tononi et al., 1998; Varela et al., 2001; Bressler and Kelso, 2001) that human cognitive function requires the emergence of globally coherent states involving the integration of specialized (segregated) information distributed throughout the cortex. Viewed in this framework, patterns of functional connectivity between brain areas arise as a result of the interplay between segregation and integration. This interplay can be captured by a global statistical measure evaluating the spectrum of statistical dependencies across all spatial scales of the system (Tononi et al., 1994; 1998). This measure captures the relative degree to which a system is simultaneously locally segregated and globally integrated. As will be discussed in greater detail below, this measure also quantifies the amount of "interesting" (non-repeating) structure present within a system's dynamics, i.e. its complexity.

To evaluate the degree to which a system's elements are statistically dependent we first define the integration of system X, consisting of n elements, as

$$I(X) = \Sigma_i H(x_i) - H(X)) \qquad (1)$$

with $H(x_i)$ denoting the entropy of the i-th element of the system considered in isolation, and $H(X)$ denoting the total entropy of the system (i.e. the joint entropy of all its constituent elements). If all i elements are statistically independent from each other, their joint entropy is equal to the sum of the individual entropies and, consequently, $I(X) = 0$. Any degree of statistical dependence (for example due to connections between the elements) will lead to a reduction of the joint entropy and produce a positive value for the integration $I(X)$ of the system. $I(X)$ will grow as the amount of statistical dependencies present within the system increases, reaching a theoretical upper bound when all elements are behaving exactly identically, in which case $I(X) = (n-1) H(x_i)$. This measure of integration (i.e. statistical dependence) can be applied to the system X in its entirety or to any of its k-out-of-n subsets of size k ($1 = k = n$). For each k, we define $\langle I(X_j^k) \rangle$ as the ensemble average of integration over all subsets of size k.

Neural complexity can be defined as the difference between the spectrum of integration obtained for a totally unstructured system of overall integration $I(X)$ (for which the spectrum of integration over subset size k is linear) and the actual ensemble averages of integration for all subset sizes 1 to n. Another, mathematically equivalent formulation gives neural complexity as the ensemble average of the mutual information between subsets of a given size (ranging from 1 to $n/2$) and their complement (Tononi et al., 1994; 1998). Thus, $C_N(X)$ is defined as

$$C_N(X) = \Sigma_k(k/n)I(X) - \langle I(X_j^k) \rangle$$
$$= \Sigma_k \langle MI(X_j^k; X - X_j^k) \rangle \qquad (2)$$

with n = number of units in system X, k = subset size, I(X) = integration of X, $I(X_j^k)$ = integration of i-th subset of size k, $\langle . \rangle$ = ensemble average over all subsets of size k, and MI = mutual information, here between a subset X_j^k of the system and its complement $X - X_j^k$.

Another, closely related (but mathematically non-equivalent) measure of complexity expresses the portion of the entropy that is accounted for by the interactions among all the components of a system (Tononi et al., 1998; Sporns and Tononi, 2002). There are three mathematically equivalent expressions for this measure, called C(X):

$$\begin{aligned} C(X) &= H(X) - \Sigma_i H(x_i | X - x_i) \\ &= \Sigma_i MI(x_i; X - x_i) - I(X) \\ &= (n-1)I(X) - n\langle I(X - x_i) \rangle \end{aligned} \qquad (3)$$

These three expressions for complexity are equivalent for all systems X, whether thpey are linear or nonlinear. Note that neither $C_N(X)$ nor $C(X)$ can take on negative values.

These measures of complexity have several interesting features:

1. Complexity is low, if all elements of a system behave alike, for example due to very strong global coupling (resulting in uniform statistical dependence) between them. In a sense, such systems are homogeneous across all spatial scales – "zooming in" on any subpart of the system yields the same integrated structure.

2. Complexity is low for systems composed of units that are statistically totally independent. For these systems, knowledge of the state of any subpart provides no information about the rest of the system (mutual information across all bipartitions within the system is zero). In the limit case of a totally random system, complexity will be zero.

3. Complexity is high, however, for systems that are internally structured such that there is high mutual information across many of its bipartitions (i.e. many of the system's subsets have mutual statistical dependencies). At different spatial scales, such systems display different kinds of dynamical relationships, for example tightly integrated activity within specialized collectives of units and weaker integrative ties between them.

In summary, complexity is low for systems that are either completely random or completely regular and ordered. In other words, for such systems, there is little we "can know" or "need to know" in order to describe their dynamics. Instead, systems that combine highly segregated (specialized) information with global integration produce high levels of mutual information within the system. In a sense, they generate high "integration of specialized information" (Tononi et al., 1998). Such systems have structured and spatially heterogeneous dynamics, i.e. different parts of the system engage in different activity patterns and yet remain highly interdependent.

It has been proposed that complex neuronal dynamics show metastability (Kelso, 1995), with systems seemingly switching between different dynamical regimes or attractor subregions and showing rich intermittency (Friston, 1997b). According to Friston (2000), brain dynamics is best described as a series of neuronal transients and dynamic instabilities, played out on complex attractor manifolds. Bressler and Kelso suggested that "metastable dynamics is distinguished by a balanced interplay of integrating and segregating influences" (Bressler and Kelso, 2001, pg. 26) and that metastability is associated with high levels of complexity. Their analysis identifies metastability and complexity as among the key origins of globally coordinated activity in the brain. Unlike dynamical regimes that are characterized by complete interdependence or randomness, metastable (complex) dynamics are inherently flexible, allowing the transient coupling of distributed cortical areas in response to differing sensory, motor and cognitive demands. Complex dynamics yields relative coordination, producing a tendency for cortical areas to form coherent patterns without becoming locked into a single persistent state. These properties of metastability are highly desirable for a cognitive architecture that is adaptive and responsive to change, while at the same time resilient with respect to external perturbations.

The precise relationship between complexity and metastability (or that between information theoretical and dynamical systems conceptualizations of complexity; see Friston 2000) remains to be elucidated. Initial theoretical analyses suggest a close association of complexity and metastability. One suggestive hint of their deep interrelatedness lies in their common structural and anatomical origins.

3. Complexity and connectivity

If complexity and metastability are indeed essential for coordination of large-scale neuronal networks in the cortex, what are some of the architectural (structural) motifs that distinguish highly complex networks from others that lack complex organizational capability? Exploratory studies using linear and nonlinear dynamics (Tononi et al., 1994; Friston et al., 1995) suggested that connectivity patterns containing a mixture of local and longe-range connections yielded highest complexity, while uniform or sparse connection patterns produced highly stereotypic or highly stochastic dynamics, both with low complexity. Given the centrality of the role of connectivity in structuring neuronal dynamics, we performed a systematic search for connection patterns giving rise to high complexity using an evolutionary optimization approach (Sporns et al., 2000a; 2000b; Sporns and Tononi, 2002). This approach, using complexity as a global cost function in optimization, allowed a relatively unbiased search within the vast space of possible connection patterns that exists for even small networks. Since complexity is a global statistical measure of functional connectivity, we essentially used functional connectivity to guide a search for specific anatomical structures in a vast space of possible networks.

Conceptualizing neuronal connection patterns as directed graphs (Sporns et al., 2000a) allowed the mathematical characterization of these patterns with a variety of rigorously defined structural measures. These measures are reviewed in detail in Sporns (2002) and Matlab functions used to calculate them are available at http://php.indiana.edu/~osporns/graphmeasures.htm. The statistical and graph theoretical analysis of networks has become a rapidly evolving field in recent years, not least due to the discovery of consistent structural principles of "small world networks" including clustering (Watts and Strogatz, 1998) and scale-free attributes (Albert and Barabasi, 2002). Numerous studies suggest that most scientifically and technologically significant large-scale networks ranging from social networks to cellular metabolism and the internet are neither random nor regular, but instead share common principles of organization (Strogatz, 2001). For the analysis of large-scale neuronal networks, several graph theoretical measures, such as the network's cluster index and its characteristic path length, are of particular interest (described in more detail below). The cluster index captures the extent to which a unit's neighbors connect to each other, forming a "clique" or local cluster. The characteristic path length is the average length of the shortest directed path between any two units in the network. The shorter the characteristic path length, the "closer" (in terms of distance in graphs), on average, are the network's units.

In extensive computer simulations (Sporns et al., 2000a; 2000b; Sporns and Tononi, 2002), we obtained consistent and highly characteristic structural patterns when optimizing for complexity, integration or entropy. Highly complex neuronal dynamics was associated with a mixture of locally dense (highly clustered) connections and long-range connections providing global integration. These networks consistently had a high cluster index combined with a relatively low characteristic path length.

To what extent are these structural motifs of "artificially" generated neuronal networks similar to those of actual cortical connection matrices? To address this question, structural measures have been applied to several large-scale cortical connection matrices (Felleman and Van Essen, 1991; Young, 1993; Scannell et al., 1999). Results indicate that the cerebral cortex is comprised of clusters of densely and reciprocally coupled cortical areas that are globally interconnected. If cortical connection matrices are "run" as dynamical systems, the resultant functional connectivity has extremely high complexity. In addition, it was found that cortical networks share some attributes of "small world architectures", including high values for cluster indices and short characteristic path lengths (Sporns et al., 2000a; Hilgetag et al., 2000). Furthermore, we found indications that cortical networks could be "wired up" using very little wiring length, a key constraint given the limited amount of volume available in real three-dimensional brains. Taken together, these studies indicated that the structural networks of the cerebral cortex possessed features that favored the emergence of complex functional dynamics.

4. Simulations of neural dynamics

Most of our previous studies employed networks with linear dynamics. To what extent do our conclusions hold for nonlinear dynamical systems? To illustrate the association of complex nonlinear neuronal dynamics with specific anatomical motifs, we simulated a simple neuronal network composed of a single map of excitatory and inhibitory units and varied the pattern of connections between excitatory units (see also Tononi et al., 1994; 1998). We recorded neural activation patterns as we varied structural connections and derived the complexity of the resulting dynamical regime. This approach is closely related to a recent study by Jirsa and Kelso on spatiotemporal pattern formation in a sheet of simulated cortical neurons with varying patterns of interconnections (Jirsa and Kelso, 2000). These authors found that variations in the connection topology lead to changes in the global dynamics of the neuronal sheet, including several spatiotemporal bifurcations.

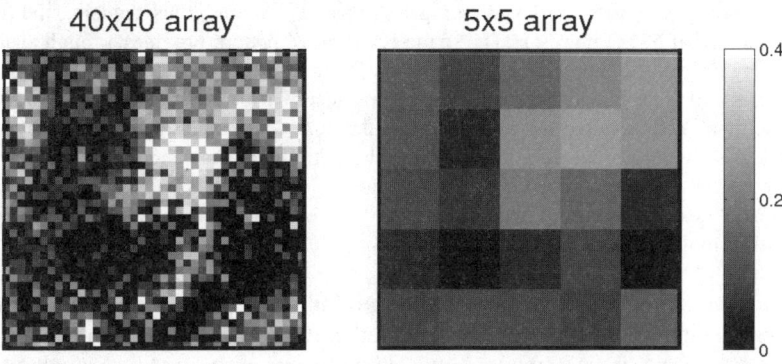

Figure 1. Activity pattern of excitatory units for the full 40×40 array (left) and the spatially averaged 5×5 array (right), for one time step of a simulation with "mixed" connection pattern. The 5×5 array is derived by averaging non-overlapping topographically mapped patches of 8×8 units in the original array.

The present model consisted of a single network containing 40×40 excitatory and 40×40 inhibitory units forming a 2-dimensional array. Excitatory and inhibitory units are modeled according to

$$s_i(t+1) = \phi[A(t) + \Omega s_i(t) + N(t)] \qquad (4)$$

where $s_i(t)$ is the activity of unit i at time t, $A(t)$ is the total synaptic input to unit i at time t, Ω is the unit's temporal persistence (0<Ω<1), $N(t)$ is Gaussian noise, and ϕ is a saturating nonlinear function, given as $\phi = \tanh(\rho[A(t) + \Omega s_i(t) + N(t)])$ if [.] > θ ($\phi = 0$ otherwise) with ρ denoting the slope of the function and θ acting as an activation threshold. A(*t*) was calculated as the linear sum of all inhibitory and excitatory inputs, i.e. as $\Sigma c_{ij}s_j(t)$. Other simulations (data not shown) involved sigmoidal nonlinearities, with essentially identical results.

At each topographic location within the map, a pair of excitatory/inhibitory units is reciprocally coupled, forming the dynamical units of the model. Each of these units is taken to correspond to a cortical column or minicolumn, with $s_i(t)$ denoting the average firing rate of its constituent excitatory and inhibitory cell populations. The dynamics of this simple minicolumn model is similar to that of classical Wilson-Cowan cortical units (Wilson and Cowan, 1973). For the parameter values used in this simulation, the elementary minicolumn unit shows spontaneous oscillatory activity.

Neuronal dynamics is generated by injecting Gaussian noise into all units of the network. No external input patterns are presented. Simulations are carried out for 21,000 iterations, and neural activations are recorded for 20,000 iterations, after an initial transient of 1,000 iterations, which is discarded. The structural network analysis utilizes the connection pattern cortico-cortical connections linking individual minicolumns across the full 40×40 array. For covariance and complexity analysis, the activity pattern of the excitatory units is spatially averaged across non-overlapping 8×8 spatial domains, thus reducing the 40×40 array to a size of 5×5 (Figure 1). The spatially averaged pattern provides a rough analogue to an array of local field potentials. Correlation matrices are obtained directly from the raw time series and are calculated for non-overlapping sets of 1000 iterations, yielding a total of 20 functional connectivity matrices per simulation. The functional dynamics of the network is stationary in all cases (discounting the initial transient) and all system variables (field potential traces) exhibit Gaussian statistics. Complexity is calculated according to Equation 3 (see above) from the entropy of the covariance (correlation) matrix, using standard formulae (Jones, 1979).

We vary the amount and spatial patterning of cortico-cortical connections linking different excitatory units across the modeled cortical array. Once connections are generated, their spatial pattern and strength remains fixed throughout the simulation. A total of four different structural patterns are examined:

"*Sparse*": No connections exist between any of the cortical minicolumns. (For this reason, structural and graph theoretical analyses are omitted for this type of connectivity.)

"*Uniform*": Connections are generated between randomly selected pairs of minicolumns, irrespective of their relative location within the map. This yields a spatially uniform pattern. Within the modeled 40×40 array, a total of approximately 20,700 connections are made, i.e. about 0.8% of all possible connections actually exist.

"*Clustered*": Connections are generated within local 5×5 neighborhoods, with a probability of 0.575, yielding a total of approximately 20,700 connections over the entire array. This connection pattern tends to connect minicolumns that are in close spatial proximity within the map.

"*Mixed*": The majority of connections are generated within local 5×5 neighborhoods, with a probability of 0.54. A few connections (with a low probability of 0.002) are made within larger neighborhoods of size 20×20. This connection pattern combines dense local connectivity with a smaller number of long-range connections linking more distant minicolumns. Long-range connections account for approximately 5% of the total.

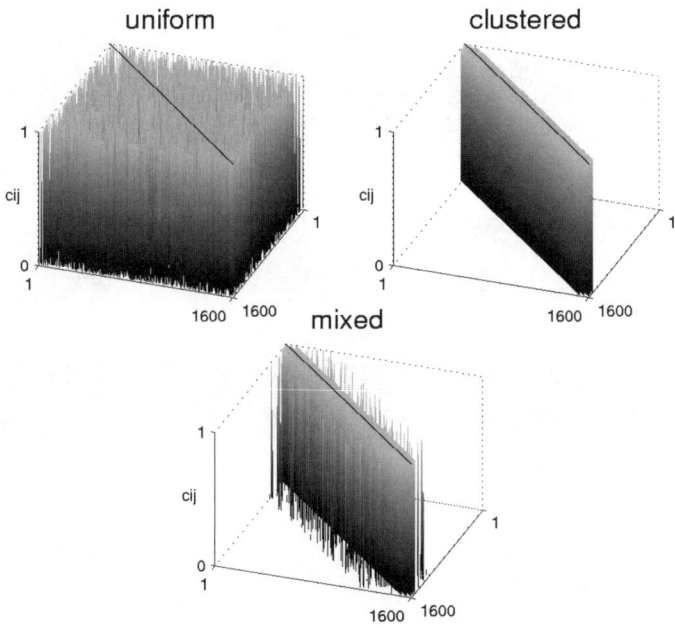

Figure 2. Connections patterns used in the computer simulations, for cases *"uniform"*, *"clustered"*, and *"mixed"* ("sparse" not shown, since no intra-map connections are generated). Plots show connection matrices ($c_{ij} = 1$, if connection is present, $c_{ij} = 0$, if connection is absent) for all 1600 excitatory units (plotted along the x- and y-axes, axis labels omitted for clarity). Note that for case *"clustered"* all connections are found close to the main diagonal of the connection matrix, since they are generated between nearby minicolumns. Case *"mixed"* additionally incorporates a number of connections between more distant excitatory units. All connection matrices contain roughly equal numbers of connections (approximately 20,200-20,700).

For all four cases, connections have fixed and uniform strength, thus normalizing the average amount of synaptic input to cortical minicolumns across different connection patterns. Also, note that in all cases approximately the same number of connections are generated (a total of ~20,200-20,700). Connection matrices for the connection patterns used in the simulations are shown in Figure 2.

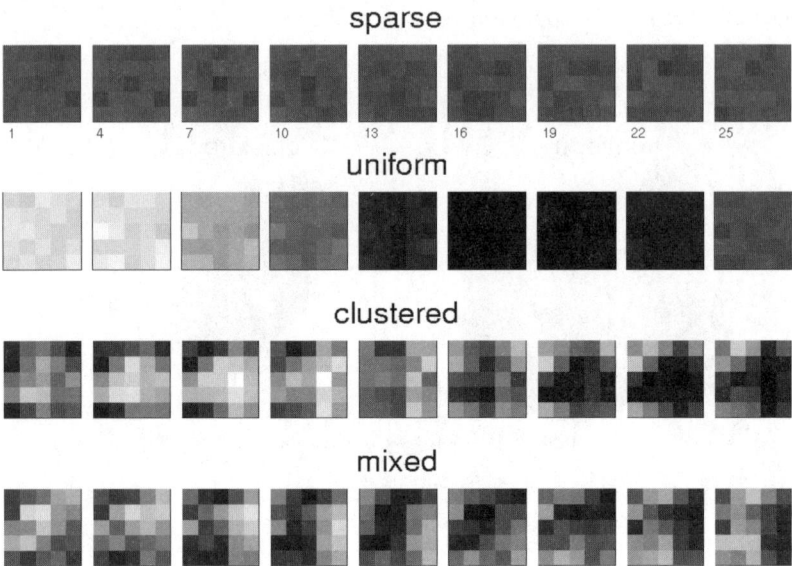

Figure 3. Sequences of activity maps from four simulations of *"sparse"*, *"uniform"*, *"clustered"* and *"mixed"* connection patterns. Individual frames show spatially averaged activity of excitatory units (gray scale plots mean activity ranging from 0 to 0.3). Frames are 3 time steps apart, spanning a total of 27 time steps, from left to right.

Figure 3 shows sequences of activity patterns obtained from four computer simulations, each with a different set of cortico-cortical connections. *"Sparse"* cortico-cortical connections yield weak activation signals due to complete desynchronization of individual cortical minicolumns. *"Uniform"* connections yield high levels of synchronization across the entire spatial extent of the simulated cortical map. *"Clustered"* connections produce strong local synchronization, with a tendency to generate "traveling waves" of excitation sweeping across the map. *"Mixed"* connections also give rise to locally synchronized activity, while traveling wave patterns are somewhat reduced, and more variable spatial patterns are found. Movies of the neuronal dynamics for these four classes of cortico-cortical connectivity can be found at http://php.indiana.edu/~osporns.

Temporal activity traces for the four connectivity patterns are shown in Figure 4. For the *"sparse"* pattern, individual minicolumns engage in oscillatory activity, but are desynchronized across the map. Thus, spatial averaging produces low amplitude and irregular activity traces. Instead, high levels of local synchrony between minicolumns as a result of dense intra-map connections (*"uniform"*, "*clustered*") are indicated by high-amplitude oscillations of the average activity trace. Waxing and waning oscillatory activity is observed for a mixture of local

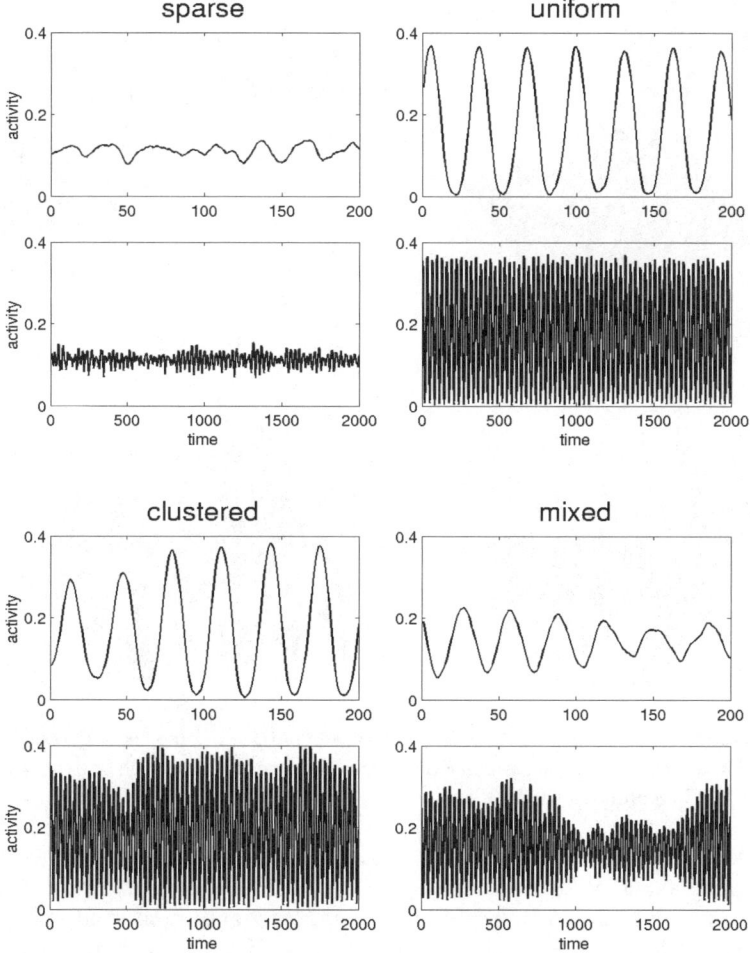

Figure 4. Temporal traces of spatially averaged excitatory activity, shown for a single 8×8 patch located near the center of the map. The plots at the top show 200 time steps, while the plots at the bottom show a total of 2000 time steps.

and long-range connections ("*mixed*"). This reflects the reduced tendency of the "*mixed*" pattern to produce traveling waves and higher local variability between minicolumns that are close to one another in the map. The power spectral density for case "*mixed*" at the dominant oscillation frequency is significantly reduced as compared to "*clustered*" or "*uniform*" (data not shown).

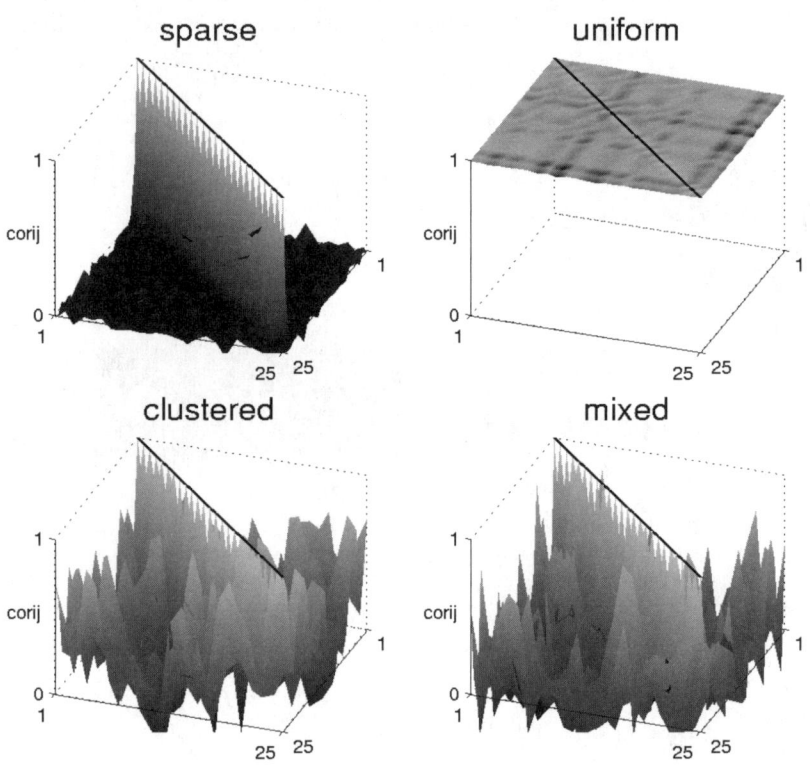

Figure 5. Correlation matrices for "*sparse*", "*uniform*", "*clustered*", and "*mixed*" connection patterns. The correlation matrix is calculated from the spatially averaged activity pattern (see Figure 1; 25 recording locations).

Figure 5 shows the correlation matrices for all four cases, obtained from recorded activity lasting a total of 20,000 time steps. The "*sparse*" pattern shows very little correlation between any of the 25 recording locations, consistent with the desynchronized appearance of the activity pattern. Conversely, the "*uniform*" pattern yields high correlation values across the entire array. Both "*clustered*" and "*mixed*" patterns produce a wide range of correlation values between different recording locations. In general, spatially proximal recording locations are highly coupled, while more distant recording locations are less strongly correlated.

Integration and complexity are measured by deriving entropies from the covariance matrices of the spatially averaged activity traces and then calculating the corresponding measures using Equations 1 and 3. Values for integration and complexity shown in Table 1 are averages derived from 20 non-overlapping periods of 1000 time steps each. Results from only one representative simulation are shown; integration and complexity values as well as structural measures are

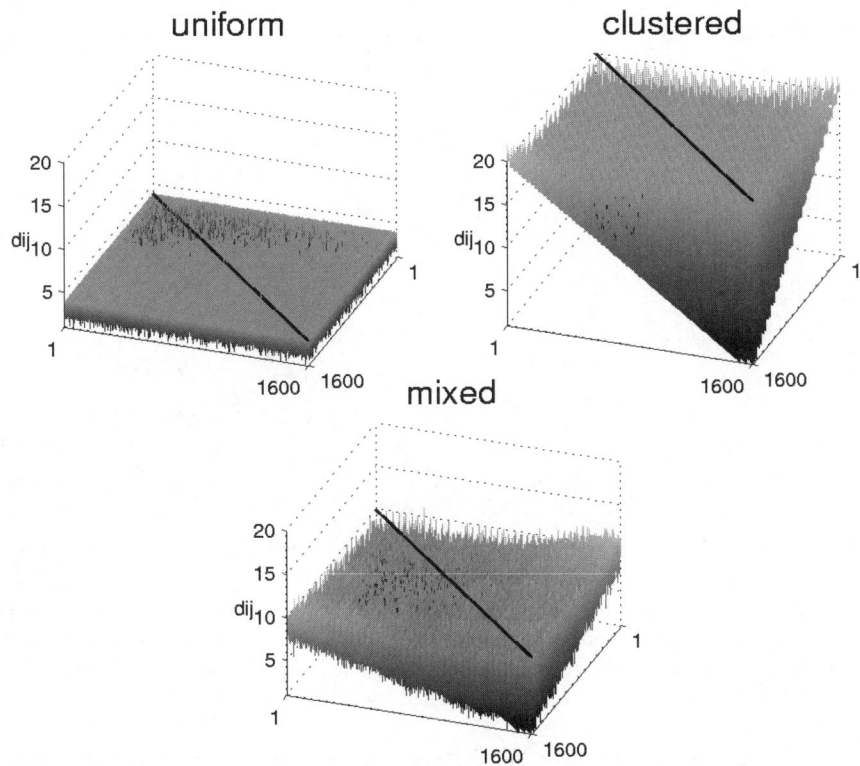

Figure 6. Distance matrices for "*uniform*", "*clustered*", and "*mixed*" connection patterns.

highly robust across multiple simulations with slightly different anatomical patterns and random noise. Integration is low for the "*sparse*" pattern (I(X) = 4.02 ± 2.17) and very high for the "*uniform*" pattern (I(X) = 69.10 ± 0.43), consistent with their overall correlation pattern (see Figure 5). Integration takes on intermediate values for both "*clustered*" and "*mixed*" connectivities (I(X) = 33.19 ± 1.34 and I(X) = 31.78 ± 2.11, respectively), reflecting the counterbalancing of global integration by formation of locally coherent but globally independent patterns. Complexity is lowest for the "*sparse*" pattern (C(X) = 0.143 ± 0.015), and is also relatively low for the pattern of "*uniform*" intra-map connections (C(X) = 0.289 ± 0.008). High values of complexity are attained for "*clustered*" connection patterns (C(X) = 0.586 ± 0.025), as well as for "*mixed*" connection patterns with a mixture of dense locally clustered connections and a small proportion of longer-range connections linking more distant minicolumns (C(X) = 0.579 ± 0.025).

Connectivity	K	I(X)	C(X)	$\lambda(G)$	$\gamma(G)$
"Sparse"	-	4.02 ± 2.17	0.143 ± 0.015	-	-
"Uniform"	20,718	69.10 ± 0.43	0.289 ± 0.008	3.1310	0.0076
"Clustered"	20,675	33.19 ± 1.34	0.586 ± 0.025	9.7418	0.3125
"Mixed"	20,196	31.78 ± 2.11	0.579 ± 0.025	5.6878	0.2637

Table 1. Total number of connections K, integration I(X) (mean ± SD of 20 consecutive time periods of 1000 time steps each), complexity C(X), characteristic path length $\lambda(G)$ and cluster index $\gamma(G)$, for one representative simulation of each of four connectivity patterns (sparse, uniform, clustered, and mixed; see text for details).

The connection patterns used in the simulations were computationally analyzed using graph theoretical methods. For each of the simulations, the corresponding connection or adjacency matrices *A(G)* were derived, each containing a total of 1600×1600 binary entries. In addition, their corresponding distance matrices *D(G)* were calculated (Sporns, 2002), shown in Figure 6. The distance matrix *D(G)* contains the distances (i.e. the lengths of the shortest paths) between any two units of the network. For example, if unit 1 is directly connected to unit 2, then d_{ij} = 1. If no path exists between the two units (i.e. unit 2 cannot be reached from unit 1), then d_{ij} = Inf. Two structural measures are shown in Table 1, the characteristic path length *λ(G)* and the cluster index *γ(G)*. The characteristic path length corresponds to the global average of the distance matrix. A low value of *λ(G)* indicated that, on average, any two units of the network are linked by short paths. The concept of distance between units in a neuronal network is significant because short distances often indicate high degrees of functional interaction, while long distances render strong interactions highly unlikely. The cluster index *γ(G)* for a network expresses the extent to which the units within the network share common neighbors that "talk" (i.e., are connected) among each other, an attribute that perhaps can be called the "cliquishness" of the network. A high cluster index points to a global organizational pattern consisting of groups of units that mutually share structural connections. Low values of *γ(G)* indicate that even locally connected units do not share many "mutual friends", i.e. are rather independent in terms of the connection patterns they maintain.

No values for $\lambda(G)$ or $\gamma(G)$ are available for the "*sparse*" connection pattern (since connections linking different minicolums are altogether absent). The "*uniform*" pattern has a characteristic path length of 3.1310, indicating that most units can be reached in very few steps. Its cluster index, however, is very low ($\gamma(G) = 0.0076$), since connections are assigned essentially at random and with low overall density. For the "*clustered*" pattern, the cluster index takes on a high value of 0.3125, since most connections are generated within local circuits. The characteristic path length is long ($\lambda(G) = 9.7418$), since units at remote locations can only be linked through numerous local (short-range) connections. In the case of "*mixed*" connections, the mixture of local and long-range connections gives rise to a combination of relatively short characteristic path length and high cluster index ($\lambda(G) = 5.6879$ and $\gamma(G) = 0.2637$). Thus, most pairs of minicolumns are only a few synaptic steps removed from each other, while at the same time remaining closely bound in local neighborhoods.

5. Outlook

Brain dynamics is shaped by the continual interplay between segregation and integration, which manifests itself as complexity and metastability across multiple spatial and temporal scales. Segregation and integration are evident in the anatomical organization of brain networks, as well as their functional connectivity recorded in the context of perceptual or cognitive processing. Segregation reflects the need to optimally extract and generate local information, while integration is necessary in order to create coherent brain states. Together, they provide the means by which the brain can (optimally?) balance the simultaneous demands of information extraction and binding, responding to the moment-by-moment challenges of the external world by selecting transient and globally coherent internal states.

Segregation and integration are functional principles that have distinct neuroanatomical origins. Throughout the thalamocortical system, there are dozens of anatomically segregated areas, linked by a network of long-range pathways. The relationship between such structural connectivity patterns and the neural dynamics they give rise to is of fundamental importance for theoretical neuroscience. Several empirical and computational approaches (Sporns et al., 2000a; Jirsa and Kelso, 2000; Friston, 2000; Bressler and Kelso, 2001) have suggested that connectivity critically determines functional interactions within local circuits as well as across extended brain systems. As more and more empirical data on neuronal connectivity patterns becomes available, it may ultimately become possible to construct a detailed connectivity matrix of a large part of the thalamocortical system, at the level of cells or minicolumns. This type of dataset could then be used to build detailed neural simulations and investigate the resulting neural dynamics, or could be subjected to graph theoretical analysis to probe for global properties of the connectivity itself.

The computer simulations presented in this chapter model the spontaneous dynamics occurring in a single cortical map, composed of minicolumn-like elements. Variations in the pattern (but not the total amount) of cortico-cortical connectivity linking these minicolumns produced marked changes in the map's functional connectivity. Consistent with earlier linear and nonlinear simulations, a pattern with *"sparse"* or *"uniform"* connectivity yielded low complexity. *"Clustered"* connectivity produced high complexity, but graph theoretical analysis revealed that this connection pattern did not allow remote minicolumns to interact along relatively direct paths. Instead, this connection pattern was associated with a high cluster index (consistent with its prevailing local connectivity) and a high value for the characteristic path length, indicating that information flow across the map needed to utilize, on average, fairly long paths. A "mixed" connection pattern, consisting of both local and long-range connections results in complex dynamics and, in addition, shows characteristics of "small-world networks". The connectivity matrix and its associated distance matrix yield high values for the cluster index and relatively low values for the characteristic path length. The combined analysis of structural and functional connectivity for these different simulated cortical maps provides additional evidence that specific classes of neural dynamics are associated with particular connection patterns (see also Sporns and Tononi, 2002). The "mixed" pattern of local and long-range connections (high complexity and small-world properties) most closely resembles the distribution of clustered and horizontal connections present in most cortical areas. More refined models and connectivity patterns are needed to investigate the differential contribution of long-range connections to dynamical patterns and to small-world attributes in more detail. Together with earlier studies of large-scale connectivity patterns (Sporns et al., 2000a; 2000b; Sporns and Tononi, 2002) the present results provide an additional indication that cortical connection patterns have small-world properties and that these properties are crucial for the generation of complex neural dynamics. It seems that small-world networks are associated with high levels of complexity as they combine locally and globally efficient information exchange (see also Latora and Marchiori, 2001). There is a growing realization in many fields of science and technology that "interesting" systems and networks are neither completely random, nor completely ordered, but rather combine elements of local order (segregation) and global interaction (integration). Brain networks seem to be no exception.

While it has been shown that highly complex networks can arise as a result of evolutionary optimization strategies, it is unknown at this time how complexity could arise in real organisms from developmental or evolutionary processes. An intriguing hypothesis involves the role of functional connectivity in generating coherent perceptual, cognitive and behavioral states. If such states could (indirectly) contribute to the evolutionary success of an organism, its brain structure might come gradually to reflect an optimal or near-optimal trade-off between local and global processing requirements, i.e. give rise to highly complex functional dynamics. According to this hypothesis, complex neural dynamics might reflect an adaptive response of brain structures to exogenous demands on rapid integration of information and coherent behavioral responses. Thus, brain complexity arises out of necessity, as a result of a dual challenge to neuronal information processing, segregation and integration.

Understanding the nonlinear dynamics of the brain is crucial for building increasingly accurate theoretical frameworks of brain function. The recent surge in studies of the statistical properties of large networks may open up new avenues for linking brain connectivity and complex neural dynamics. How network structure gives rise to the wide range of dynamical phenomena observed in real nervous systems is a question that will occupy Coordination Dynamics for many years to come.

References

Albert R, Barabasi AL (2002) Statistical mechanics of complex networks. Rev Mod Phys 74, 47-97

Bressler SL, Coppola R, Nakamura R (1993) Episodic multiregional cortical coherence at multiple frequencies during visual task performance. Nature 366, 153-156

Bressler SL (1995) Large-scale cortical networks and cognition. Brain Res Rev 20, 288-304

Bressler SL, Kelso JAS (2001) Cortical coordination dynamics and cognition. Trends Cogn Sci 5, 26-36

deCharms RC, Zador A (2000) Neural representation and the cortical code. Annu Rev Neurosci 23, 613-647

Ding M, Bressler SL, Yang W, Liang H (2000) Short window spectral analysis of cortical event-related potentials by Adaptive MultiVariate AutoRegressive (AMVAR) modeling: Data preprocessing, model validation, and variability assessment by bootstrapping. Biol Cybern 83, 35-45

Felleman DJ, Van Essen DC (1991) Distributed hierarchical processing in the primate cerebral cortex. Cereb Cortex 1, 1-47

Frackowiak RSJ, Friston KJ, Frith CD, Dolan RJ, and Mazziotta JC (1997) Human Brain Function. Academic Press, San Diego, CA

Friston KJ (1994) Functional and effective connectivity in neuroimaging: A synthesis. Hum Brain Mapp 2, 56-78

Friston KJ, Tononi G, Sporns O, Edelman GM (1995) Characterizing the complexity of neural interactions. Hum Brain Mapp3, 302-314

Friston KJ (1997) Imaging cognitive anatomy. Trends Cogn Sci 1, 21-27

Friston KJ (1997) Transients, metastability, and neuronal dynamics.Neuroimage 5, 164-171

Friston KJ (2000) The labile brain. I. Neuronal transients and nonlinear coupling. P Roy Soc Lond B Bio 355, 215-236

Hilgetag CC, Burns GAPC, O'Neill MA, Scannell JW, Young MP (2000) Anatomical connectivity defines the organization of clusters of cortical areas in the macaque monkey and the cat. Philos T Roy Soc B 355, 91-110

Jirsa VK, Kelso JAS (2000) Spatiotemporal pattern formation in neural systems with heterogeneous connection topologies. Phys Rev E 62, 8462-8465

Jirsa VK, Kelso JAS (2003) Integration and segregation of perceptual and motor behavior. This volume

Jones DS (1979) Elementary Information Theory. Clarendon Press, Oxford, UK

Kelso JAS (1995) Dynamic Patterns. MIT Press, Cambridge, MA

Kötter R (2001) Neuroscience databases: Tools for exploring brain structure-function relationships. Philo T Roy Soc B 356, 1111-1120

Kötter R (2002) Neuroscience Databases. A Practical Guide. Kluwer Academic Publishers, Boston, MA

Latora V, Marchiori M (2001) Efficient behavior of small-world networks. Phys Rev Lett 87, 198701

McIntosh AR, Nyberg L, Bookstein FL, Tulving E (1997) Differential functional connectivity of prefrontal and medial temporal cortices during episodic memory retrieval. Hum Brain Mapp 5, 323-327

McIntosh AR (1999) Mapping cognition to the brain through neural interactions. Memory 7, 523-548

McIntosh AR, Rajah MN, and Lobaugh, NJ (1999) Interactions of prefrontal cortex related to awareness in sensory learning. Science 284, 1531-1533

Mesulam MM (1998) From sensation to cognition. Brain 121, 1013-1052

Roelfsema PR, Engel AK, König P, Singer W (1997) Visuomotor integration is associated with zero time-lag synchronization among cortical areas. Nature 385, 157-161

Scannell JW, Burns GAPC, Hilgetag CC, O'Neil MA, Young MP (1999) The connectional organization of the cortico-thalamic system of the cat. Cereb Cortex 9, 277-299

Singer W, Gray CM (1995) Visual feature integration and the temporal correlation hypothesis. Annu Rev Neurosci 18, 555-586

Sporns O, Tononi G, Edelman GM (1991) Modeling perceptual grouping and Figure-ground segregation by means of active reentrant circuits. P Natl Acad Sci USA 88, 129-133

Sporns O, Tononi G, Edelman GM (2000a) Theoretical neuroanatomy: Relating anatomical and functional connectivity in graphs and cortical connection matrices. Cereb Cortex 10, 127-141

Sporns O, Tononi G, Edelman G (2000b) Connectivity and complexity: the relationship between neuroanatomy and brain dynamics. Neural Networks 13, 909-922

Sporns O, Tononi G (2002) Classes of network connectivity and dynamics.Complexity 7, 28-38

Sporns O (2002) Graph theory methods for the analysis of neural connection patterns. In: Kötter R (ed.) Neuroscience Databases. A Practical Guide. Kluwer, Boston, MA, 169-183

Srinivasan R, Russell DP, Edelman GM, Tononi G (1999) Increased Synchronization of Neuromagnetic Responses during Conscious Perception. J Neurosci 19, 5435-5448

Strogatz SH (2001) Exploring complex networks. Nature 410, 268-277

Tononi G, Edelman GM, Sporns O (1998) Complexity and coherency: Integrating information in the brain. Trends Cogn Sci 2, 474-484

Tononi G, Sporns O, Edelman GM (1994) A measure for brain complexity: Relating functional segregation and integration in the nervous system. P Natl Acad Sci USA 91, 5033-5037

Varela F, Lachaux JP, Rodriguez E, Martinerie J (2001) The brainweb: Phase synchronization and large-scale integration. Nat Rev Neurosci 2, 229-239

von Stein A, Rappelsberger P, Sarnthein J, Petsche H (1999)Synchronization between temporal and parietal cortex during multimodal object processing in man. Cereb Cortex 9, 137-150

Ward LM (2003) Oscillations and synchrony in cognition. This volume

Watts DJ, Strogatz SH (1998) Collective dynamics of 'small-world' networks. Nature 393, 440-442

Wilson HR, Cowan JD (1973) A mathematical theory of the functional dynamics of cortical and thalamic nervous tissue. Kybernetik 13, 55-80

Young MP (1993)The organization of neural systems in the primate cerebral cortex. P Roy Soc Lond B Bio 252, 13-18

Oscillations and Synchrony in Cognition

Lawrence M. Ward

University of British Columbia

"Relax," said the doorman, "we are programmed to receive." - The Eagles, Hotel California

1. Introduction

This chapter reviews work on five themes relevant to the role of coordination dynamics in cognition. It is to be seen as a complement to the intriguing review of Bressler and Kelso (2001) and therefore will not repeat any of that material, although convergence is inevitable and will be apparent to readers of both. First, I will review the properties of relaxation oscillators, which I believe form a useful class of models for the dynamic aspects of cognition. Then I will discuss the possible role(s) of such oscillators in short-term memory, psychophysical judgment, attention, and consciousness. The approach of this chapter is, admittedly, highly speculative. Many of the ideas are only half-baked, some are still sticky dough with some crucial ingredients still to be mixed in, a few are only recipes that might not even prove to be edible. Nonetheless I believe that the juxtaposition of these ideas does lead in some interesting directions (to mix metaphors if not bread dough). Thus, I won't apologize further for them, or indicate their respective states of doneness; I leave that to the reader.

The points I make in this chapter are intended to apply more or less equally both to cognitive processes and to the brain processes that presumably underlie them. It is possible that some models of cognitive processes can be built up from those of underlying brain processes, or can at least be constrained by those models, and vice versa. At the least, a close relationship between cognitive and brain processes is assumed by several modern disciplines, especially that of cognitive neuroscience, which is the paradigm that guided the present review. Moreover, within that paradigm, I take an interactive, parallel dynamical view, rather than a sequential processing view. We can think of the sequential view as a sort of pinball machine of the brain, in which information about a stimulus enters a sensory system and is transmitted from one brain center or cognitive process to another, "lighting up" each in turn, until it is "processed" and thereby causes a response. Figure 1a shows a typical model of such processing. Although there is a dynamic in Figure 1a, for example information does "flow" along various pathways, it is a relatively uninteresting flow, and not sufficient to explain many of the phenomena we wish to explain (e.g., conscious awareness: see Lamme, 2000). The approach I will take is more interactive: it stresses information flow along numerous parallel pathways between large numbers of brain areas or cognitive processes. These information flows *oscillate*, and so do the "stimuli" with which they interact. Thus, the response of the brain or cognitive system to a stimulus should be modeled as oscillations interacting with oscillations, changing them thereby, and thus generating the ongoing rhythm of behavior. One interesting metaphor for this is Foucault's Pendulum, which, through its interactions with the rotation (rotary oscillation) of the earth, enabled Foucault to demonstrate that motion. Figure 1b shows an illustrative picture of the brain viewed from this perspective. Notice that, although somewhat fictional, this picture has multiple pathways between areas and many possibilities for complex oscillations. All of these interactions, of course, would be critically affected by the timing of the information flow along the pathways. I will be concentrating especially, in this chapter, on simultaneity, that is, on events that happen at the same time in different parts of the cognitive or brain system.

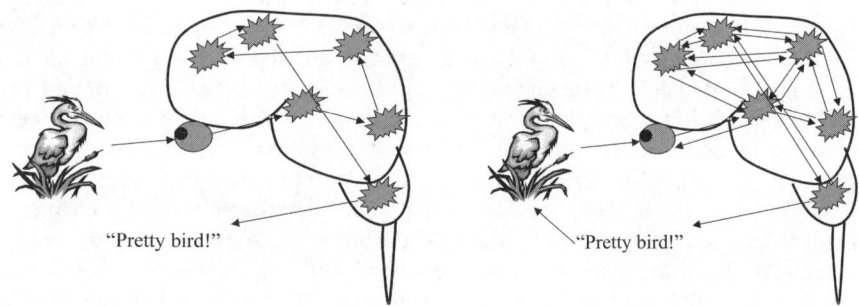

Figure 1 (a) A "pinball" model.   **(b)** A "pendulum" model.

2. Relaxation oscillators

Possibly the most useful oscillators relevant to cognition, and certainly one of the most interesting, is the relaxation oscillator. The prototypical relaxation oscillator is defined by the following nonlinear, second-order differential equation (van der Pol, 1926):

$$\frac{d^2v}{dt^2}+\alpha(v^2-1)\frac{dv}{dt}+\omega^2v=0 \tag{1}$$

Van der Pol (1926) studied this equation in an analog electrical circuit containing a neon lamp in parallel with a capacitor, these elements being placed in series with a resistor and a battery. The variable v stands for the voltage across the neon lamp. In Equation 1 the middle term on the left side describes the damping force, which works in opposition to the restoring force, $\omega^2 v$, overwhelming it when v becomes large and causing the oscillations. The relationship between a and ω determines the behavior of the oscillator. When $a \ll \omega$ (Figure 2, top), the oscillator behaves like a linear oscillator with a very small amount of damping, generating a very slowly decaying sine-wave-like oscillation. When $a = \omega$ (Figure 2, middle), the oscillation becomes less like a sine wave but is still regular. When, however, $a \gg \omega$ (Figure 2, bottom), the oscillation resembles a square wave. In each cycle, sudden transitions (nearly vertical traces) are preceded and followed by relatively slow, small-amplitude, voltage changes.

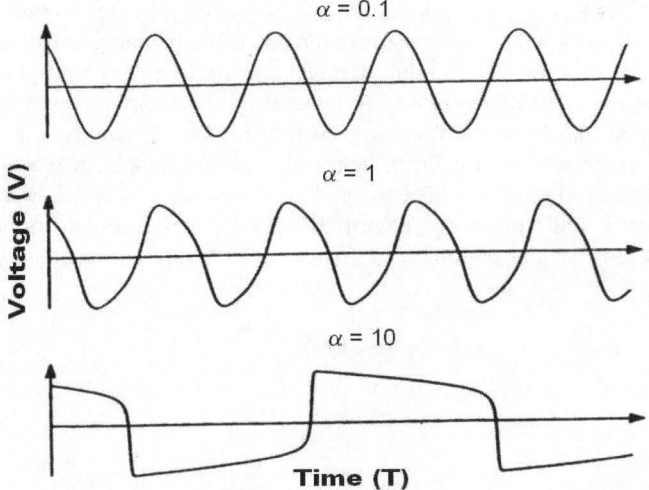

Figure 2. Behavior of van der Pol relaxation oscillator, $\omega = 1$ (after van der Pol, 1926).

Van der Pol (1926) labeled the latter case a "relaxation oscillation" because it follows the slow buildup and fast release (relaxation) of charge in a capacitor. In this system the sudden relaxation is the important characteristic of the oscillation, not the slow buildup of the restoring force. The most relevant feature of relaxation oscillators is that they easily achieve synchrony with a forcing input while maintaining relatively constant amplitude.

FitzHugh (1961) noticed that the spiking behavior of neurons closely resembles relaxation oscillations: the action potential is like the release of charge from a capacitor, and the relatively slower buildup of voltage before the next action potential, usually under the influence of input from other neurons, is like the buildup of charge in a capacitor. He simplified the Hodgkin and Huxley (1952) model of this process in the squid giant axon, using Equation 1 with $\omega = 1$ plus some additional terms, and showed how it mimicked the behavior of the Hodgkin-Huxley equations. The resulting equations, also studied by Nagumo, Arimoto and Yoshizawa (1962), are called the FitzHugh-Nagumo model of the neuron. The FitzHugh-Nagumo equations are:

$$\frac{dv}{dt} = \alpha(w + v - \frac{v^3}{3} + z) \qquad (2)$$

$$\frac{dw}{dt} = -(v - c + dw)/\alpha \qquad (3)$$

where v represents the fast, spiking process, w represents the slower, recovery process, z represents input from other neurons, a is the damping constant (greater than $\omega = 1$ so that the system shows relaxation oscillations), and c and d are constants that affect the recovery rate. The FitzHugh-Nagumo model of the neuron is the model of choice in biophysics, probably because it simply and elegantly captures the essence of neural firing behavior. In many biophysical applications, stochastic (noise) forcing is added to the "fast" Equation 2, and z is eliminated from Equation 2 and replaced by deterministic, usually sinusoidal, forcing in the "slow" Equation 3 (e.g., Longtin, 1993). This results in a pair of equations

$$\varepsilon \frac{dv}{dt} = v(v-a)(1-v) - w + \xi(t) \qquad (4)$$

$$\frac{dw}{dt} = v - cw - b - r\sin(\beta t) \qquad (5)$$

Longtin (1993) showed that a numerical solution to Equations 4-5 displays neuron-like behavior, generating action-potential-like oscillations in single spikes and bursts. Furthermore, as expected because they are a type of relaxation oscillator they synchronize to periodic forcing (Longtin, 1995a) and exhibit stochastic phase locking (Longtin, 1995b). Figure 3 shows a spike train generated by a numerical solution to Equations 4-5.

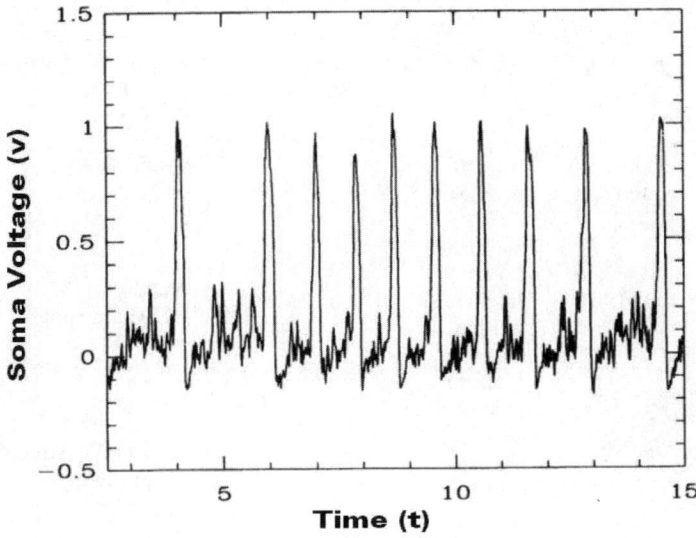

Figure 3. Spike train generated by Equations 4-5 (after Longtin, 1993)

3. Hierarchies of relaxation oscillators

Given the attractive properties of relaxation oscillators, and particularly their ability to capture important aspects of neuron dynamics, it is tempting to speculate that they might provide a basis for linking neural and cognitive processes. One problem with making this linkage is that brain processes and cognitive processes take place at different time scales. The main idea of this section is that relaxation oscillators can be used to model the dynamics at multiple time scales and thus help to provide the required linkage. There are many scales of time and space in the brain-mind system, but it is useful to distinguish at least three of them.

Table 1 summarizes these three scales. The fastest time scale, that of individual neurons or small groups of neurons, can be modeled directly with the version of relaxation oscillators illustrated by FitzHugh-Nagumo or other neuron models (Equations 4 and 5). Larger groups of neurons (e.g., those performing within - cortical-area computations) and their associated cognitive processes require relaxation oscillators with somewhat different oscillation properties,

Time Scale (ms)	Cycle freq (Hz)	Neuron System	Consciousness Status	Perceptual/Cognitive Task
1 to 2.5	1000 to 400	1 to 10 neurons	Preconscious	Peripheral sensory processing
10 to 25	100 to 40	10s to 100s of neurons: within area circuit, local synchrony	Unconscious	Sternberg (1970) STM scanning
100 to 250	10 to 4	1000s to millions of neurons: between area loops, global synchrony	Conscious	Crovitz (1970) strobe matching

Table 1. Some time scales of the mind-brain system (after Ward, 2002a).

corresponding to, for example, the 40-Hz oscillations discovered by Gray and Singer (1989). These would summarize the actions of coupled systems of neuron-like oscillators. At the highest level, characterized by between-cortical-area loops, for example, yet other oscillators would be required to model the global oscillations of the brain (EEG) and mind (conscious awareness), representing the interactions of coupled oscillators at the intermediate level.

Although at present this program is highly speculative, there is some reason to believe that it is worth pursuing. First, the three time/space scales described in Table 1 have some claim to usefulness. Consider their ontological status. For example, Ward (1988) described the ontological boundary between unconscious and conscious processing in terms of several behavioral phenomena that define it. First, above this boundary subjects report that they "know" the correct response in a psychophysical task, whereas below it they report they are only guessing (although $d' > 0$ indicates that their responses convey information about the stimulus - see Merikle & Cheeseman, 1986). Second, Stroop color-word priming effects are observed only if a word can be perceived with $d' > 0$, but the color-word stimulus does not have to be consciously perceived. Only when the color-word stimulus is consciously perceived, however, is such priming affected by the proportion of congruent primes relative to incongruent primes. Third, lexical decisions are primed by both meanings of polysemous words when such a prime is not consciously perceived, but by only the meaning associated with a previously presented word when the prime is consciously perceived. Finally, The phenomena of memory scanning and strobe light matching described in Table 1 (and in more detail in section 7) also have different properties above and below this boundary. Second, I have already described the intimate relationship between relaxation oscillators and neurons. And relaxation oscillators are beginning to appear in models for some cognitive processes, such as those involved in music cognition (e.g., Eck, 2002). Finally, relaxation oscillators also appear in models of cortical dynamics, such as those attempting to explain the origin of the electroencephalogram (EEG). A particularly intriguing example of the latter is Nunez's (e.g., 2000) dynamical theory of the relationship between brain activity and EEG recordings. Among other things, the theory assumes that neural networks operate in a background of oscillatory changes in the number of active synapses, and that cognitive and overt behavior depend on the levels of excitatory and inhibitory synaptic action density (number of active synapses per unit brain volume), and action potential density, in various brain regions. Nunez (2000) also assumed that the multiple scales of dynamics described by the theory can be summarized by two prominent scales of oscillations and their interaction. He described *local oscillations* in synaptic action density as the result of interactions between neurons in relatively small functionally segregated areas of the brain, connected in neural networks through both excitatory and inhibitory connections. Such oscillations could be taken, for example, to represent the response of a sensory system to a stimulus. These correspond to the middle level described in Table 1. *Global oscillations*, on the other hand, result from interactions between local regions. Such oscillations could be taken to represent "higher" processing of sensory stimuli, and resemble the highest level described in Table 1. Nunez (2000) derived various simplified expressions for local and global oscillations and their interaction under some plausible assumptions about brain structure and function. One result was a set of five second-order differential equations, each one a type of relaxation oscillator as in Equation 6:

$$\frac{d^2Z}{dt^2} - \varepsilon(1-Z^2)\frac{dZ}{dt} + [k_n^2 - (1+\varepsilon)Z^2]Z = 0 \qquad (6)$$

Here Z stands for the synaptic action density (like voltage in the van der Pol Equation 1), e is the damping parameter, and $[k_n^2 - (1+e)Z^2]Z$ is the restoring force term. Numerical solutions to one set of such equations (with nonzero right-hand side containing the parameter q) exhibit both local (faster) and global (slower) oscillations that resemble prominent EEG waves such as theta and gamma waves, which are associated with information processing by the brain. Figure 4A shows these nested oscillations in simulation results from Nunez (2000) and Figure 4B shows similar nested oscillations in the human cortex. Clearly, not only do such oscillations occur in the human brain, but also a form of relaxation oscillator can capture their prominent qualitative features.

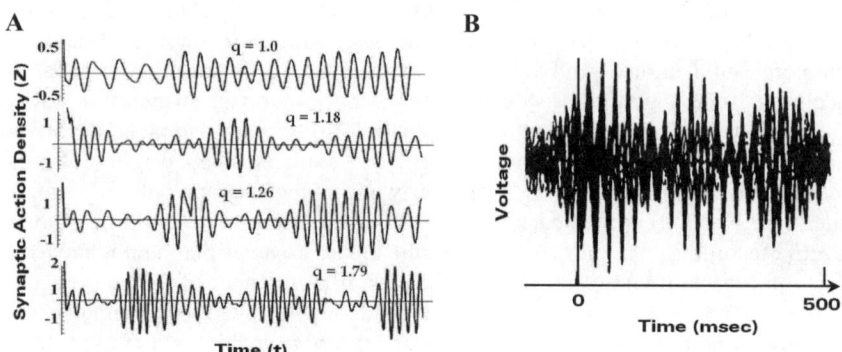

Figure 4. (A) Global and local oscillations in Nunez's model of EEG (after Nunez, 2000). (B) Nested oscillations in a magnetoencephalographic recording of human cortical activity (after Lisman & Idiart, 1995)

4. Short term memory

What role could such nested oscillations play in cognition? One possibility is dramatized by a model of "short-term" memory proposed by Lisman and Idiart (1995). They proposed that gamma (approximately 40 Hz) oscillations nested within slower theta (approximately 6-7 Hz) oscillations reflect the processes that give rise to temporary memory for information just experienced. It is well-known that such short-term memory can hold approximately 7±2 "chunks" of material, whether it be letters, words, numerals, telephone numbers, etc. (Miller, 1956; Simon, 1974). Lisman and Idiart (1995) studied a network of groups of (model) neurons oscillating at the theta frequency, each group spatially encoding a single alphabetic letter to be "remembered." The network proved to be capable of maintaining the firing and correct phases of up to seven groups of neurons, each active during a different subcycle of the theta-frequency oscillation. Figure 5 shows how this occurs in the model. In this theory, one "rehearsal" of all short-term memory "slots" takes place during every (global) theta cycle, rehearsal of each item taking place at the (local) gamma frequency, about 40 Hz or one item every 25 msec, the maximum rate at which short-term memory can be "scanned" (Sternberg, 1970). Therefore, about 7±2 chunks of information (≈40/6) can be retained in short-term, memory. Notice in the bottom row of traces in Figure 5 that when a new letter ("X") is introduced, encoded by a separate group of neurons from those encoding the current contents of memory, one of the previously remembered letters ("R") is no longer remembered (indicated by the absence of a spike on the rightmost half of the "R" trace). The network can only store 7 chunks, and if a new chunk is introduced one of the older ones must be "forgotten."

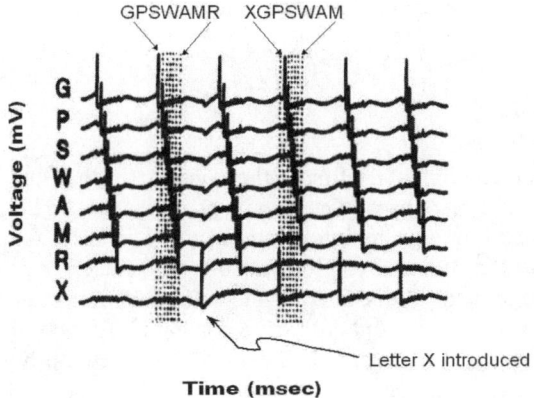

Figure 5. Short-term memory encoding in oscillatory subcycles (after Lisman & Idiart, 1995)

5. Psychophysical scaling

Relaxation oscillators also could play a role in explaining some of the complexities found in the data of psychophysical scaling studies. For many years, in order to obtain clean and meaningful data, psychophysicists have asked experimental subjects to produce time series of judgments. For example, a typical experiment in magnitude estimation, in which observers produce numerical responses proportional to their experience of sensation magnitude, involves a series of several hundred judgments made over many minutes, perhaps several hours. These psychophysical scaling judgments display several interesting and, in the measurement context, somewhat distressing, properties. For example, repetitions of the same stimulus intensity do not result in identical responses; on the contrary, responses to identical stimuli can vary widely. Moreover, identical responses are given to quite different stimulus intensities. Finally, these time series of scaling judgments display complex temporal structures, called "sequential dependencies." These are usually modeled empirically either with a multiple regression on the logarithms of current and previous stimuli and responses, or by a combination of a multiple regression on current and previous stimuli with an ARIMA model of response dependencies. Equation 7 is a typical multiple regression equation accounting for from 60% to 80% of the variance in time series of scaling judgments:

$$\log ME_n = \beta_0 + \beta_1 \log S_n + \beta_2 \log ME_{n-1} + \beta_3 \log S_{n-1} + \varepsilon \qquad (7)$$

The coefficient β_1 in Equation 7 is taken to be an estimate of the exponent, m, in the psychophysical power law,

$$\overline{ME} = a(S - S_0)^m \tag{8}$$

whereas the other coefficients estimate the constant in that law or the effects of previous stimuli and responses. The exponent of the lower law is an important fundamental variable in psychophysics and sensory science, for it is supposed to represent the transfer function of the sensory transducer (e.g., Stevens, 1975). Indeed, for average data produced from the same laboratory, the exponent varies systematically across sensory modalities in a meaningful way that can be related to the other properties of the modalities, such as their ability to discriminate intensity differences, and also to characteristics of their underlying physiology. For example, the average exponent found by Stevens (1975) for magnitude estimates of the brightness of a disk-shaped light source is about 0.3, whereas that for magnitude estimates of the loudness of a 1000-Hz tone is about 0.6. Unfortunately, even when care is taken to remove variance arising from sequential dependencies, estimates of the exponent in Equation 8 are not stable across individuals, across laboratories, or even across similar experiments in the same laboratories. The exponent differences are too large to have arisen from transducer differences, and they swamp the differences in average exponents. For example, Table 2 shows some exponents estimated from magnitude estimations of the

Subject	Exponent	R^2
1	0.71	0.88
2	0.33	0.83
3	0.75	0.62
4	1.00	0.80
5	0.34	0.66
6	0.71	0.75
M	0.64	
SD	0.26	

Table 2. Exponents of the psychophysical power law (Equation 8) for loudness (data from West, Ward & Khosla, 2000)

loudness of a 1000-Hz tone made by six different individuals. In this case, the average exponent value of 0.64 is not too far off the canonical value of 0.6 favored by Stevens (e.g., 1975), but the individual valuesvary from 0.33 to 1.00.

Many researchers have tried to explain such results, but there is little agreement, and several have suggested that we simply give up trying to measure such psychological variables as loudness, brightness, saltiness, and so forth, and concentrate instead on understanding how sensory systems evolved (e.g., Lockhead, 1992). While endorsing the effort to understand the evolution of sensory systems, I am building a dynamical model that might help explain the above results as well as suggest a remedy that would rescue psychophysical measurement from its difficulties (Ward, 2002b). The model represents a sensory stimulus by (locally) synchronous firing in a group of (sensory) neurons. Consistent with the approach of Nunez (2000) described earlier, I take the synaptic action density associated with the representing sensory neurons to be the macroscopic field variable that represents the magnitude of the sensory stimulus. For example, a pure tone excites a group of neurons in primary auditory cortex that includes those tuned to the tone's frequency, and usually many others at nearby frequencies; the more intense the tone, the more neurons are firing. The total number of active synapses in this group of neurons represents the tone's intensity. However, this local, modality-specific, encoding of stimulus intensity must be mapped to a neural representation of a number in the magnitude estimation scaling procedure, or to that of a stimulus in another modality in a technique called cross-modality matching. One way to do this would be by bringing a sensory representation into synchrony with a central neural process that would recruit neurons into the process until the sensory input was matched, probably via other oscillating networks. The synaptic action density of the matching network would be a modality-independent representation of the psychological magnitude of the sensory stimulus, abstracted from its peripheral representation. A report of this magnitude could be made either by adjusting input from another modality, say light, until the same central magnitude was achieved (cross-modality matching), or by recruiting other, cognitive representations, say of numbers or of categories, that would in turn control verbal or key-press responses (magnitude estimation).

This suggestion resembles that of Ward (1991) regarding a modality-independent representation of psychological magnitude. Figure 6 shows a diagram of how such a system would work, where the central process is labeled *CO*. I have begun to study how to represent all of the processes involved as relaxation oscillators at various levels. Of course, the neurons representing the stimulus are themselves such oscillators. They combine to produce a middle-level (local) oscillator whose amplitude represents the experienced sensory magnitude. Another, (local) central oscillator, *CO,* must match the sensory oscillator both in period and phase and in amplitude. Its amplitude, then, becomes the modality-free representation of sensory magnitude that can be mapped into numbers or other sensory magnitudes via other central or peripheral oscillators. I have begun by using a modification of the FitzHugh-Nagumo equations to model the sensory oscillator. Euler (numerical

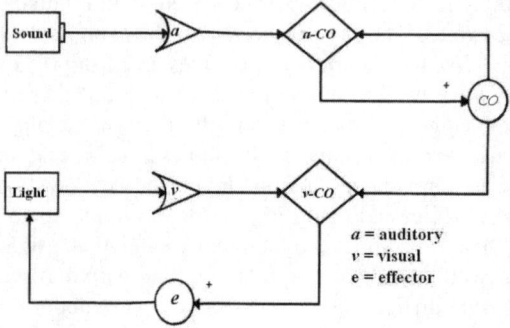

Figure 6. Model of cross-modality matching (after Ward, 1991)

integration) solutions to these equations with time step Δt yields:

$$v_t = v_{t-1} + \alpha\left(w_{t-1} + s - \frac{v^3}{3} + v_{t-1}\right)\Delta t \qquad (9)$$

$$w_t = w_{t-1} - \Delta t\left(\frac{v_{t-1} - a + bw_{t-1}}{\alpha}\right) + \xi(t) \qquad (10)$$

where v is sensory magnitude, s is stimulus intensity in dB, and $\xi(t)$ is Gaussian noise. An interesting property of these equations is revealed in Figure 7: what appears to be a chaotic attractor. So, for reasonable values of the parameters in Equations 9-10 we see that the modeled representation of the stimulus is chaotic.

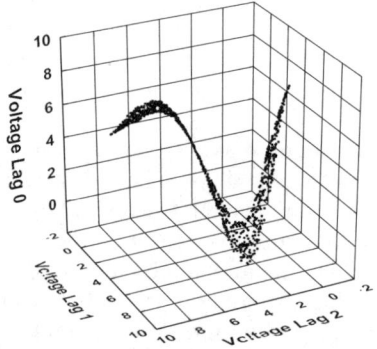

Figure 7. Chaos in the stochastic FitzHugh-Nagumo oscillator.

This means that sensory magnitude cannot be a constant over time, even for a "constant" stimulus, which immediately explains several of the properties of scaling judgments described earlier. Such chaos makes the representation of the stimulus "fuzzy" as asserted by Ward (1979) in the paper that describes a theory based on the empirical regression model in Equation 7. This theory explains the sequential dependencies as a consequence of this fuzziness. In the present theory these would arise from the more global interactions of the various local oscillators, which are described briefly in Ward (2002b). Finally, although we can't get rid of such oscillations, since they are fundamental to the perceptual and cognitive processes involved, we can create a context in which they are acknowledged and, to the extent possible, controlled. West, Ward and Khosla (2000) describe our attempt to do this.

6. Attentional oscillations

Everyone knows what attention is. It is the brain's selective processing mechanism and it oscillates over time and space. Yes, that is a slight updating of William James's famous definition of attention ("Everyone knows what attention is. It is the taking possession by the mind, in clear and vivid form, of one out of what seem several simultaneously possible objects or trains of thought." James, 1890, I, 403-404). Although everyone agrees that attention changes over space and time (see, e.g., Sperling and Weichselgartner, 1995), very few consider modeling those changes for longer than a second or so. Two exceptions are the model of predator vigilance by Dukas and Clark (1995), and the model of attentional entrainment to rhythmical events proposed by Large and Jones (1999). I will discuss each of these briefly as stepping stones to a hoped-for more general model of attentional oscillations based on relaxation oscillators.

6.1. Predator vigilance

Figure 8 summarizes the Dukas and Clark (1995) model of the foraging behavior of an animal. It is assumed that foraging requires attentional effort (vigilance) and that available attentional resources decline exponentially with time since foraging began. When resources are too low to sustain foraging any longer, the animal rests, during which resources are replenished, again exponentially. These processes are modeled by a simple linear differential equation, which gives the curves shown in Figure 8. The model captures several important aspects of animals' foraging behavior but it is incomplete in several respects. First, Figure 8 shows only one foraging cycle. An animal's entire lifetime could be modeled by repeating the foraging cycle many times, possibly with some variation in the parameters of the exponential curves as the animal became ill, recovered, was injured, recovered, and aged. Second, each cycle is deterministic. From laboratory studies of vigilance in humans (for example watching a radar screen for the signal of an ICBM) it is known that vigilance is highly variable even within an overall fatigue-recovery cycle. This variability could be modeled as a stochastic component added to the Dukas and Clark model, or it might

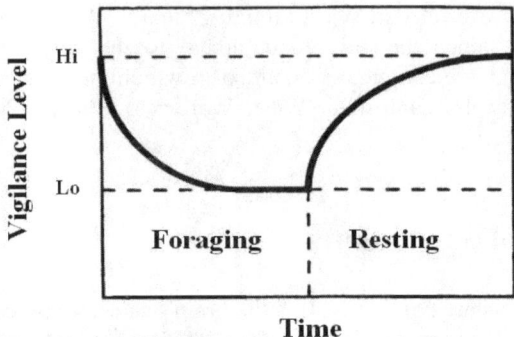

Figure 8. A model of predator vigilance (after Dukas & Clark, 1995).

indicate that more complex dynamics are at work. Ward and Richard (Ward, 2002a) found that fluctuations of simple reaction time in a vigilance task had a $1/f^a$ power spectrum, with $a \approx -0.6$ and decreasing as decision load increased. This could indicate equally-weighted summing of fluctuations at several time scales (see Ward, 2002a), or coupling of those fluctuations (Greenwood & Ward, 2002), or some other source of $1/f$-like noise. Finally, the exponential form of vigilance decline and recovery, although plausible, has not been documented. Indeed, published graphs of decline in sensitivity to rare signals show a variety of forms, some resembling those of relaxation oscillators (see Figure 2). Thus, it is possible that a more general model of vigilance might be found in equations such as Equation 1, the van der Pol relaxation oscillator.

6.2. Attending to rhythmic events

A more directly oscillatory model of attention has been proposed by Large & Jones (1999). Their model addresses how attention can be coordinated with external events that occur at particular moments in time, particularly those that are part of a rhythmical structure such as music. In their model internal "pulses" of attention come to be coordinated with external events through entrainment of an internal oscillator, both in relative phase, ϕ_n, and in period, p_n. Attentional oscillations in the Large and Jones (1999) model have the following structure:

$$\phi_{n+1} = \phi_n + \frac{t_{n+1} - t_n}{p} - \eta_\phi F(\phi_n) \qquad (\mathrm{mod}_{-0.5, 0.5} 1) \qquad (11)$$

$$p_{n+1} = p_n + p\eta_p F(\phi_n) \qquad (12)$$

$$F(\phi) = \frac{1}{2\pi} \sin 2\pi\phi \qquad (13)$$

Equation 13 provides the (sinusoidal) oscillations, whereas Equations 11 and 12 represent the process of entrainment of those oscillations to external events occurring at times t_n and t_{n+1}. In Equation 11, $(\mathrm{mod}_{-0.5,0.5}1)$ means that relative phase varies only between -0.5 and 0.5, and η_f is the entrainment parameter for relative phase. In Equation 12, η_p is the entrainment parameter for period, and p (no subscript) represents the period of the external rhythm. The entrainment parameters determine how quickly the relative phase and period, respectively, of the attentional oscillator approach those of the external rhythm. The model also assumes that the more closely attention is coordinated with the occurrence of the relevant external events, the more tightly focused in time are the attention pulses. This process is described by another equation in which the extent of the temporal distribution of attention is a monotonic function of the difference in phase between the attentional oscillator and the external rhythm.

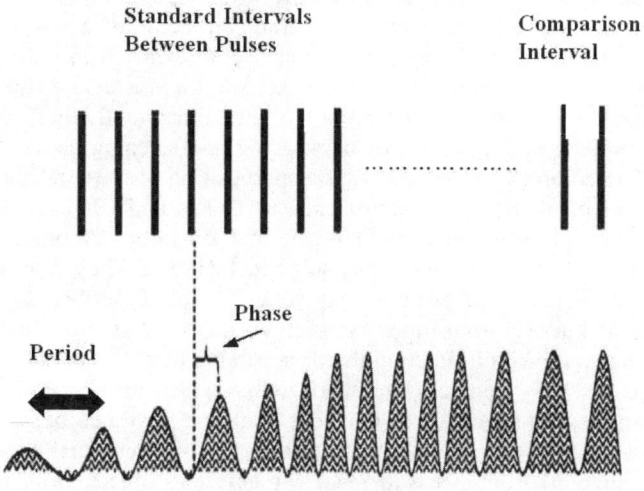

Figure 9. Entrainment of an attentional oscillator (after Barnes & Jones, 2000).

Figure 9 shows pictorially how the attentional oscillator becomes entrained with external events in a time judgment task. Broad, uncoordinated attentional pulses occur at the beginning of the regular sequence of external events, say tones, because synchronization is poor initially. As the entrainment progresses, through Equations 11-13, the period and phase of the attentional oscillator become more like those of the sequence of tones, and the pulses themselves become more tightly focused in time as synchronization improves. In Figure 9, the subject's task is to say whether the comparison inter-onset interval (IOI) created by the two isolated tones is the same, shorter than, or longer than the standard IOI created by the last two tones of the regular sequence. Clearly the longer the entraining sequence, the more aligned will be the attentional oscillator with the relevant tone onsets, and the more focused attentional resources will be on the relevant moments of time, and so the more accurate the temporal judgments will be (the smaller the difference threshold for interval). This prediction was confirmed in several studies by Large and Jones (1999). More recently, variations of time judgment performance with the duration and rate of entraining events were also predicted by this model (Barnes & Jones, 2000), as were variations in pitch judgment accuracy with the timing of the target tones relative the entraining sequence (Jones, Moynihan, MacKenzie & Puente, 2002).

6.3. Spontaneous oscillations in modality, space, and time

Unfortunately, important environmental events do not always occur in strictly rhythmical sequences so that the oscillator described by Large and Jones (1999) can be entrained to them and muster resources optimally. Important events are distributed in modality, space and time. And yet there is a rhythm to their distribution, since we can anticipate many of them from the temporal contingencies created by their causal structure. And there is also a rhythm to our behavior, created by the waxing and waning of attentional (and other) resources in the cycle of preying and sleeping, or of working and sleeping as more of us do these days. The complex dynamics of the superposition of these rhythms has not yet been modeled, although the approaches of Dukas and Clark (1995) and of Large and Jones (1999) represent steps in that direction. Another promising approach is that of Sperling and Weichselgartner (1995). They represent visual attention as a sequence of discrete "episodes," each described by a spatial distribution function. The transitions between episodes are assumed to be smooth and governed by a transition function that is separable from the spatial distribution function. This is a very general framework within which many theories could be cast. For example, the sequence of episodes could be governed by an oscillator, perhaps of relaxation type, that caused transitions in attentional focus (spatial distribution function) from place to place, or object to object, in a rhythm that depended on the task at hand. To accomplish this, the parameters of the oscillator could be functions of other variables, such as arousal, threat, goal, and so forth.

Whatever the form of the theory that attempts to describe the oscillations of attention in space and time, it is certain that there will be several time and space scales involved. Thus, since all cognitive processes are noisy, it is likely that fluctuations of attention will display the signature of multi-scale complexity, a power spectrum of so-called $1/f^a$ character. The power spectrum summarizes how much sinusoidal variation at each of several small bands of frequency is contained in a time series (see Ward, 2002a, for a more complete discussion, or a text on the Fourier transform for even more). Systems that produce noise at several scales and

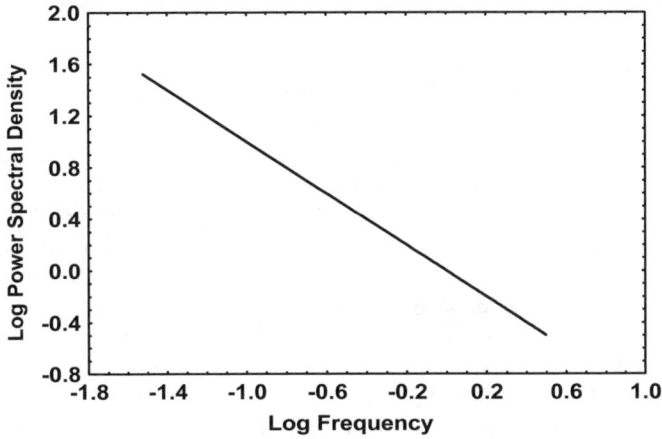

Figure 10. Power spectrum of 1/f noise.

that have no preferred scale (one whose noise dominates) produce noise whose power spectrum resembles the plot in Figure 10. In that Figure, log power spectral density is plotted against log frequency, so that the power function $psd = 1/f^a$ becomes $\log psd = -a \log f$, the equation of a straight line. In Figure 10, the value of a is 1, and the noise producing this power spectrum is called "$1/f$ noise" or "pink noise." This kind of noise has been taken to be a signature of complex processes such as self-organization. It does indeed occur in many sequences of cognitive acts, such as reaction times in classification, search, memory scanning, and other cognitive tasks (e.g., Gilden, 2001, Ward, 2002a).

7. Synchrony and consciousness

Nowhere do oscillations and synchrony seem as exotic as in the study of consciousness. Ever since Gray and Singer (1989) discovered synchronous oscillations at around the gamma frequency (40 Hz) in the visual cortex of cats researchers have considered the possibility that there is something special about neurons firing in synchrony. The most prominent suggestions have been that synchronous neural firing is responsible for binding together the disparate groups of neurons that encode the various properties of sensory stimuli, such as color, shape, motion, depth, and spatial locus of visual objects. These neurons are distributed in different layers and pathways of the visual system, and yet there seems to be no central group of neurons that represents the integration of all of these properties into the solid, stable, full-featured *gestalt* that we experience. The solution to this "binding" problem could be that the neurons whose firing represents the various features of an object come to fire in synchrony, whereas neurons representing features of other objects are firing asynchronously with those neurons (but possibly synchronously among themselves). Crick and Koch (1990) suggested that this 40-Hz synchronous neural firing might be the neural correlate of visual awareness.

More recent studies have linked synchronous oscillations of neural activity ever more closely with conscious awareness of sensory stimuli. For example, changes in conscious awareness of one or the other of two binocularly rivaling visual stimuli are accompanied by a change in the synchrony of the firing of the various neurons representing the stimuli. There is widespread coherence between the magnetoencephalogram (MEG) at various non-sensory brain sites and the MEG of sensory neurons responding to a stimulus that is currently in consciousness, whereas there is no such coherence for stimuli present on the retina but currently not in consciousness ("suppressed" in terms of binocular rivalry; Tononi, Srinivasan, Russell & Edelman, 1998). Coherence is related to the square of the correlation coefficient between two time series, in this case of MEG measurements. This coherence suggests that the neurons in these areas are firing synchronously, even though they are separated by several synapses. Rodriguez, George, Lachaux, Martinerie, Renault, and Varela (1999) recorded the electroencephalogram (EEG) while subjects viewed an ambiguous visual stimulus that could be perceived as either a face or as a meaningless shape. They found that when subjects reported seeing a face there was a pattern of phase synchronization of the EEG at the gamma frequency across widely separated brain areas that did not appear when a meaningless pattern was reported. In a different context, Miltner, Braun, Arnold, Witte and Taub (1999) found that awareness of the signaling properties of a stimulus in an associative learning experiment was correlated with long-range synchrony at the gamma frequency. A possible basis for these findings is the suggestion of Lamme (e.g., 2000) that feedforward processing in any neural system is not sufficient for conscious awareness. Feedback, or reentrant, processing is also necessary, forming recurrent loops among neural populations that could be the basis for their coming into synchrony. There surely seems to be something special about synchronous neural activity.

7.1. The cortical dynamic core

Tononi and Edelman (1998) suggested that cortical neural activity that achieves both *differentiation* (selection from a large number of possible states) and *integration* (widespread coherence resulting in a particular one of those states) is necessary for conscious experience. Both are achieved by their concept of the *dynamic core*, that is, "...a large cluster of neuronal groups that together constitute, on a time scale of hundreds of milliseconds, a unified neural process of high complexity..." (Tononi & Edelman, 1998, p. 1849). The cortical dynamic core integrates parallel processing in the brain in consciousness, with various aspects of the current sensory, cognitive, emotional, and motor processing making up the dynamic core, and thus being "in consciousness," while other aspects of simultaneously occurring processing are outside the dynamic core, and thus are outside of consciousness. Neural synchrony, possibly arising from the global oscillations described by Nunez (2000), across widespread areas of the brain would bind neural activity in disparate regions of the brain into the dynamic core, with unsynchronized activity continuing outside of awareness. A large-scale (65,000 integrate-and-fire neurons, similar to FitzHugh-Nagumo neurons, and 5 million connections) simulation of cortico-cortico and thalamo-cortical circuits in the brain demonstrated the plausibility of this idea (Lumer, Edelman & Tononi, 1997). When the simulated neural network was stimulated with continuous input, multilevel synchronous oscillations at roughly 50 Hz emerged. These oscillations were not programmed into the simulation in any way, they simply occurred as a consequence of the interactions between the neural activity in the various simulated brain areas in response to the stimulus input.

The involvement of relaxation oscillators in this process should be clear. As described above, neurons behave as relaxation oscillators. And model neurons, such as the FitzHugh-Nagumo, display the tendency of relaxation oscillators to phase lock with periodic forcing while maintaining constant amplitude (Longtin, 1995a, 1995b). Networks of such model neurons have been studied extensively, and the conditions under which they spontaneously synchronize have begun to be understood. Two recent papers are of particular interest in this connection. First, Bose and Kunec (2001) studied a pair of mutually coupled self-inhibitory Hodgkin-Huxley (Hodgkin & Huxley, 1952) type model neurons with variable synaptic delay applied to the coupling. They showed that such pairs exhibit transient synchronous oscillations arising from post-inhibitory rebound. This synchrony depends on long synaptic delays, so it could apply to cortical neurons that interact through several synapses. The synchronization happens because one neuron's spiking behavior moves from one nullcline to another on the phase plane, thus "catching up" to another neuron relative to which its phase is lagging. A nullcline is a trajectory in phase space along which a dynamic system moves when the time derivative of one of the dependent variables is 0. Fox, Jayaprakash, Wang and Campbell (2001) studied one- and two-dimensional arrays of relaxation oscillators very similar to the FitzHugh-Nagumo model neuron, with excitatory coupling between nearest neighbors in the arrays and conduction delays applying to this coupling. For pairs of neurons, synchronization occurred within a cycle or two for a cubic y-nullcline (trajectories where $dy/dt = 0$). For the one- and two-dimensional arrays, synchronization time depended somewhat on the number of oscillators and their connectivity, but was achieved for most arrays. It was very fast for small one-dimensional arrays, and did not vary significantly with either choice of y-nullcline or conduction delay. For larger one-dimensional arrays, and two-dimensional arrays, convergence to synchrony was slower, sometimes requiring 6-12 periods or more. Again, synchronization is achieved in these systems by movement between nullclines caused by interaction between neurons via their coupling, in this case excitation transferred from a neuron to its neighbors, or received by a neuron from its neighbors. And this system synchronizes in spite of conduction delays between the neurons, even inhomogeneous delays. Thus, synchronization is a stable solution for a wide range of systems of coupled relaxation oscillators, both model-neuron-like and also other types. This means that synchronization could occur between relaxation oscillators at several scales of time and space, as required for cognitive operations in the brain. Thus, it is plausible that the synchrony displayed by the cortical neurons in the cortical dynamic core of Tononi and Edelman (1998) arises from the properties of coupled relaxation oscillators. This synchrony could arise spontaneously in the cortico-cortico circuits, but work such as the simulations of Lumer et al. (1997) implies that the thalamo-cortical circuits must also be involved (see Ward, 2003). In particular, it has been suggested that the nucleus reticularis of the thalamus, which wraps around the dorsal thalamus like a layer of an onion, might modulate the firing of thalamic neurons, and thus help achieve synchrony among cortical neurons (e.g., Barth & MacDonald, 1996; see also Sherman & Guillery, 2001).

7.2 A thalamic dynamic core?

Given the possible role of the thalamus in achieving synchrony among cortical neurons, it is useful to explore further how this part of the brain might be involved in the synchronous oscillations that appear to be the neural correlate of conscious awareness. In this section I will briefly summarize the argument of Ward (2003) in favor of a more central role for the thalamus than that of simply mediating cortical synchrony. In that paper I suggest that the synchronous dynamic core that creates conscious awareness is actually located in the thalamus. This argument depends on three major points: (1) normal activity in the thalamus is necessary for conscious awareness; (2) the roles of the cortico-cortico circuits and the cortico-thalamic circuits seem to be calculating and displaying, respectively; and (3) we are aware of the products of psychological processes, not of the processes themselves.

The first point is established by several experiments involving general anesthetics that abolish conscious awareness (Alkire, Haier, & Fallon, 2000; John, 2001). In particular, Alkire et al. (2000) used positron emission tomography (PET) to establish that the thalamus is one of only two brain areas that lie in the intersection of the action of two very different general anesthetics. The other is the brainstem reticular formation, usually credited with providing non-specific arousal through diffuse pathways to the rest of the brain during the waking state. Alkire et al (2000) proposed that a hyperpolarization block of neurons in the relay nuclei of the dorsal thalamus is the common mechanism of unconsciousness in all general anesthesia. This implies that normal firing of these thalamic neurons is critical for normal consciousness. A hyperpolarization block would not only suppress firing in these neurons, it would also disrupt their ability to fire synchronously with each other and with cortical neurons. By itself, this finding cannot distinguish between the possibility that the thalamus induces cortical synchrony and the possibility that the thalamus is the locus of the dynamic core (coincidentally also inducing cortical synchrony). Thus, the second and third points are also necessary to establish a case for the thalamic dynamic core.

The second point has to do with the putative functions of the cortico-cortico and cortico-thalamic circuits. The circuits are summarized in Figure 11. All cortical areas that project to other cortical areas also receive projections back from those areas (subcortical areas too; e.g., Zeki, 1993). The functions of these circuits were addressed in two elegant papers by Mumford (1991, 1992). He argued that interacting pairs of cortical areas have a similar relationship whatever the nature of the information they are processing. Each pair has a "higher" and "lower" member; the higher member receives the results of the calculations performed by the lower pair as "data" and performs its own calculations on these data, sending the results back down to the lower area in the form of "hypotheses." It also sends these results to the next higher area, to which it is a lower area, as data. Eventually the each pair of interacting areas comes to a form of adaptive resonance (e.g., Grossberg, 2000). Thus, each pair of interacting cortical areas forms a loop of recurrent processing; many such loops, some connecting across large cortical distances, make up the cortical circuits.

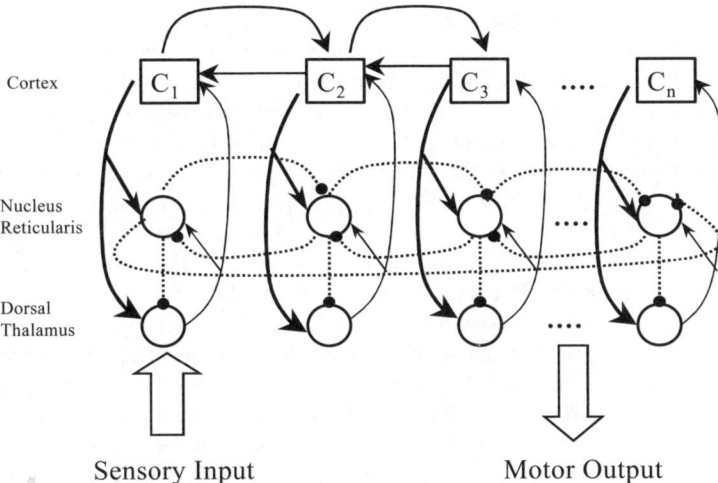

Figure 11. Cortico-thalamic and thalamo-cortical circuitry. Dotted lines with dots are inhibitory, solid lines with arrows are excitatory. (Ward, 2003)

The function of the thalamo-cortical circuits is more problematic. It is clear that some thalamic nuclei relay sensory information to cortical areas. Many analogous nuclei in the dorsal thalamus, however, have no sensory input whatsoever. Moreover, there are massive projections from cortex to thalamus for each of the relay nuclei, whether they have sensory input or not. These projections are several times as dense as those from the thalamus to the cortex, even for sensory relay nuclei. It might be the case that cortical projections modify the firing of the sensory relay nuclei relative to the calculations they and their communicating cortical areas are performing. However, what could the function of the non-sensory (or "secondary") relay nuclei be? There is nothing coming in to modify. Mumford (1991) suggested that the function of all of the relay nuclei in the thalamus is that of "active blackboards" for the calculations performed by their associated cortical areas. For the sensory nuclei, the relationship would be much like that between cortical areas: the results of the calculations performed by the associated cortical area would be presented as hypotheses and the interaction between incoming data and hypotheses would settle into a stable interpretation. For the non-sensory nuclei, the results of cortical calculations would simply be displayed in the associated thalamic nuclei and then represented to the cortex to be incorporated in its next round of calculations. In both cases, the relay nuclei would function as active blackboards, displaying the results of previous cortical calculations for 100 or so milliseconds. In the thalamic dynamic core proposal, some but not all of these displays would be integrated into a dynamic core of synchronous neural activity via the inhibitory interactions between the associated areas of the nucleus reticularis (see Figure 11). Notice in Figure 11 that these areas also receive inputs both from the cortical areas and from the relay nuclei, but their inhibitory projections only go to the relay nuclei. Of course, since each relay nucleus is linked to a particular cortical area, synchrony among groups of neurons in the relay nuclei would give rise to synchrony among groups of neurons in the associated cortical areas, which can be measured by MEG or EEG.

The story just told about the possible functions of cortico-cortico and cortico-thalamic circuits is interesting, but it too does not suffice to establish the active blackboards/relay nuclei as the locus of a dynamic core of conscious awareness. The third point mentioned earlier is also necessary: we are conscious of the products of the cortical calculations, not of the calculations themselves. There seems to be consensus that the cortex calculates the contents of conscious awareness (e.g., Rees, Kreiman & Koch, 2002). That we do not experience the details of the calculations also seems to be a generally accepted tenet of cognitive psychology. A somewhat anecdotal example is memory retrieval: we are aware of memories, not of the "search" process. Nor are we aware of the codes used to accomplish the search. When we can't "find" a memory, we simply try to think of related things, not to modify the memory code so that the memory is retrievable. When asked whether we know something, like a telephone number, we can say whether we know it or not by "looking" where in memory it would be and either finding something there or not. Interestingly, we can often find "something" in that place but not be able to retrieve it. What is this looking process? We can't say. Indeed, memory researchers spend a lot of time trying to characterize it objectively, without much help from phenomenology. Another, more technical, example concerning memory, involves scanning short-term memory for the presence of particular items, like letters or numerals. When given a set of, e.g., numerals to remember in short term memory and then asked whether a particular target numeral is in this memory set, people typically respond quickly either yes or no. Their response time depends linearly on the size of the memory set: it increases about 25-40 msec with each additional numeral in the set, indicating that subjects are serially searching the memory set for the target digit (some additional data is necessary to distinguish this from parallel search in which search efficiency declines with increasing set size). Importantly, the slopes of the linear functions relating response time to memory set size are identical for yes and no answers for subjects who are moderately practiced at the task, indicating that they are doing a serial exhaustive search of the memory set in both cases (Sternberg, 1970). This is completely at odds with subjects' experience of doing this task. Typically, people feel that when they perform this task the answer simply "pops" into their heads. They feel they are doing a parallel search, or possibly a serial self-terminating search, since when the target numeral is in the memory set they might find it before they have checked all numerals in the set. No one ever experiences a serial exhaustive search; indeed most people feel that it would be inefficient and "stupid" to do such a search. They are astonished when they see the data. People can be asked to do a serial search consciously; they report that it feels quite different - much slower and certainly self-terminating. This is consistent with an observation of Crovitz (1970) that people can consciously follow a strobe light's oscillations only up to a rate of about 4 to 5 Hz (250 to 200 msec per cycle). After that, although they can resolve the flashes up to the critical flicker frequency, they experience it as a flickering light, not as a series of independent flashes. More examples could be adduced, for example creativity, the illusion of free will (Crick, 1994; Wegner, 2002), nearly all perceptual processing, and so forth, but I believe the point is made: we consciously experience products, not processes.

Thus, it is possible that the synchronous dynamic core that mediates conscious awareness is actually located in the thalamus. Normal thalamic activity is selectively blocked by general anesthetics whose action is to abolish conscious awareness. Cortical activity still occurs during the unconscious period, although, of course it is probably not normal either, especially if the thalamus mediates its synchronous firing. The cortex appears to perform complex calculations and send the results to other cortical areas and to the thalamic relay nuclei, which could act as active blackboards for the associated cortical areas. Finally, since we experience the results of cortical calculations and not the calculations themselves, and these results are sent to the thalamic relay nuclei via massive projections, perhaps the display of those results in the thalamus constitutes conscious awareness of them. Central in this proposal is the role of synchronous activity of relaxation oscillators (neurons). There is something special about such activity; it even seems to be responsible for consciousness when implemented in a biological brain.

References

Alkire MT, Haier RJ, Fallon JH (2000) Toward a unified theory of narcosis: Brain imaging evidence for a thalamocortical switch as the neurophysiologic basis of anesthetic-induced unconsciousness. Conscious Cogn 9, 370-386

Barnes R, Jones MR (2000) Expectancy, attention and time. Cognitive Psychol 41, 254-311

Barth DS, MacDonald KD (1996) Thalamic modulation of high-frequency oscillating potentials in auditory cortex. Nature 383, 78-81

Bose A, Kunec S (2001) Synchrony and frequency regulation by synaptic delay in networks of self-inhibiting neurons. Neurocomputing 38-40, 505-513

Bressler SL, Kelso JAS (2001) Cortical coordination dynamics and cognition. Trends Cogn Sci 5, 26-36

Crick F (1994) The Astonishing Hypothesis: The Scientific Search for the Soul. Simon & Schuster, London

Crick F, Koch C (1990) Some reflections on visual awareness. Cold Spring Harb Sym 55, 953-962

Crovitz F (1970) Galton's walk: Methods for the analysis of thinking, intelligence and creativity. Harper & Row, New York

Dukas R, Clark CW (1995) Sustained vigilance and animal performance. Anim Behav 49, 1259-1267

Eck D (2002) Finding downbeats with a relaxation oscillator. Psychol Res 66, 18-25

FitzHugh RA (1961) Impulses and physiological states in theoretical models of nerve membrane. Biophys J 1, 445-466

Fox JJ, Jayaprakash C, Wang D, Campbell SR (2001) Synchronization in relaxation oscillator networks with conduction delays. Neural Comput 13, 1003-1021

Gray CM, Singer W, (1989) Stimulus-specific neuronal oscillations in orientation columns of cat visual cortex. P Natl Acad Sci USA 91, 6339-6343

Greenwood PE, Ward LM (2002) Aggregated autoregressive systems with 1/f properties. UBC, Unpublished ms

Grossberg S (2000) The complementary brain: Unifying brain dynamics and modularity. Trends Cogn Sci 4, 233-246

Hodgkin AL, Huxley AF (1952) A quantitative description of membrane current and its application to conduction and excitation in nerve. J Physiol 117, 500-544

James W (1890) The Principles of Psychology. Henry Holt and Company, NY

John ER (2001) A field theory of consciousness. Conscious Cogn 10, 184-213

Jones MR, Moynihan H, MacKenzie N, Puente J (2002) Temporal aspects of stimulus-driven attending in dynamic arrays. Psychol Sci 13, 313-319

Lamme VAF (2000) Neural mechanisms of visual awareness: A linking proposition. Brain Mind 1, 385-406

Large EW, Jones MR (1999) The dynamics of attending: How people track time-varying events. Psychol Rev 106, 119-159

Lisman JE, Idiart MAP (1995) Storage of 7±2 short-term memories in oscillatory subcycles. Science 267, 1512-1515

Lockhead GR (1992) Psychophysical scaling: Judgments of attributes or objects? Behav Brain Sci 15, 543-601

Longtin A (1993) Stochastic resonance in neuron models. J Stat Phys 70, 309-327

Longtin A (1995a) Synchronization of the stochastic FitzHugh-Nagumo equations to periodic forcing. Il Nuovo Cimento 17D, 835-846

Longtin A (1995b) Mechanisms of stochastic phase locking. Chaos 5, 209-215

Lumer ED, Edelman GM, Tononi G (1997) Neural dynamics in a model of the thalamocortical system I. Layers, loops and the emergence of fast synchronous rhythms. Cereb Cortex 7, 207-227

Merikle PM, Cheeseman J (1986) Consciousness is a "subjective" state. Behav Brain Sci 9, 42-43

Miller GA (1956) The magical number seven, plus or minus two: Some limits on our capacity for processing information. Psychol Rev 63, 81-97

Miltner WHR, Braun C, Arnold M, Witte H, Taub E (1999) Coherence of gamma-band EEG activity as a basis for associative learning. Nature 397, 434-436

Mumford D (1991) On the computational architecture of the neocortex I. The role of the thalamo-cortical loop. Biol Cybern 65, 135-145

Mumford D (1992) On the computational architecture of the neocortex II. The role of cortico-cortical loops. Biol Cybern 66, 241-251

Nagumo J, Arimoto S, Yoshizawa S (1962) An active pulse transmission line simulating nerve axon. P IRE 50, 2061-2070

Nunez PL (2000) Toward a quantitative description of large-scale neocortical dynamic function and EEG. Behav Brain Sci 23, 371-437

Rees G, Kreiman G, Koch C (2002) Neural correlates of consciousness in humans. Nat Rev Neurosci 3, 261-270

Rodriguez E, George N, Lachaux JP, Martinerie J, Renault B, Varela FJ (1999) Perception's shadow: Long-distance synchronization of human brain activity. Nature 397, 430-433

Sherman SM, Guillery RW (2001) Exploring the Thalamus. Academic Press, San Diego

Simon HA (1974) How big is a chunk? Science 183, 482-488

Sperling G, Weichselgartner E (1995) Episodic theory of the dynamics of spatial attention. Psychol Rev 102, 503-532

Sternberg S (1970) Memory-scanning: Mental processes revealed by reaction-time experiments. In: Antrobus JS (ed.) Cognition and Affect. Little, Brown and Company, Boston, 13-58

Stevens SS (1975) Psychophysics: Introduction to its Perceptual, Neural, and Social Prospects. Wiley Interscience, New York

Tononi G, Edelman GM (1998) Consciousness and complexity. Science 282, 1846-1851

Tononi G, Srivinivasan R, Russell DP, Edelman GM (1998) Investigating neural correlates of conscious perception by frequency-tagged neuromagnetic responses. P Natl Acad Sci USA 95, 3198-3203

van der Pol B (1926) On "relaxation-oscillations." Phil Mag J Sci 2, 978-992

Ward LM (1988) At the threshold of fundamental psychophysics. In: Ross JS (ed.) Fechner Day 88. International Society for Psychophysics, Stirling, Scotland, 17-22

Ward LM (1979) Stimulus information and sequential dependencies in magnitude estimation and cross-modality matching. J Exp Psychol Human 5, 444-459

Ward LM (1991) The measurement of Ψ. In: S Bolanowski, G Gescheider (eds) Measurement of psychological magnitude. Lawrence Erlbaum, Hillsdale, NJ

Ward LM (2002a) Dynamical Cognitive Science. MIT Press, Cambridge, MA

Ward LM (2002b) Synchronous relaxation oscillators and inner psychophysics. In: da Silva JA, Matsushima EH, Ribeiro-Filho NP (eds.) Fechner Day 2002. International Society for Psychophysics, Rio de Janeiro, 145-150

Ward LM (2003) A New Cartesian Theater of Consciousness. Unpublished ms, University of British Columbia

Wegner D (2002) The Illusion of Conscious Will. MIT Press, Cambridge, MA

West RL, Ward LM, Khosla R (2000) Constrained scaling: The effect of learned psychophysical scales on idiosyncratic response bias. Percept Psychophys 62, 137-151

Zeki S (1993) A Vision of the Brain. Blackwell Scientific Publications, London

Integration and Segregation of Perceptual and Motor Behavior

Viktor K. Jirsa & J. A. Scott Kelso

Center for Complex Systems & Brain Sciences, Florida Atlantic University, Boca Raton FL33431

We are developing a conceptual framework that provides a general basis for the dynamic grouping of individual components. Perceptual grouping is the process by which raw elements are aggregated into larger and more meaningful collections (Feldman, 1999). The emergence and disappearance of such a group has been termed differently in different areas of Science and Philosophy, such as integration and segregation, convergence and divergence, binding and loss thereof. We wish to widen the notion of grouping by viewing motor processes and perceptual processes induced by sensory information as equivalent events over time defined in their appropriate spaces (Kelso et al., 1990). Many operational formalisms for the treatment of the temporal relationship of typically two or four coupled sensorimotor components have been developed within the field of coordination dynamics (see Kelso, 1995 for a review). Within the domain of visual (Feldman, 1999), auditory (Bregman, 1990) and multi-sensory perception (Stein & Meredith, 1993), numerous systematic studies have been performed identifying parameters and conditions under which the formation of percepts changes. Less frequently, the interdependence of percept formation and motor trajectory formation has been studied (Bogaerts et al., 2003). In the current chapter we wish to identify the mutual features and factors determining percept and motor trajectory formation. Here grouping can impose decisive influences on other low-level processes resulting in different percepts, e.g. lightness perception (Gilchrist, 1977), and different movement patterns, e.g. reduced reaction times (Davis, 1959).

In the following we will discuss general features of grouping in various sensory and motor domains. We will then show that spatiotemporal cohesion in grouping may be understood as phase locking in rhythmic paradigms and, more generally, extended to discrete and transient behaviors through the notion of trajectory convergence.

1. Grouping in the perceptual and motor domain

When temporal sequences of sensory stimuli are presented to human subjects, their perception depends on a set of control parameters, which typically either alter the timing properties, such as the inter-stimulus-intervals (ISIs), or qualities of the stimulus, such as volume and pitch, or brightness and color. The stimulus qualities have been the focus of many discussions, primarily in the visual domain, in the context of the organization into striate patterns (Barchilon Ben-Av & Sagi, 1995; Kubovy & Wagemans, 1995, Zucker et al., 1983;), contours (Caelli & Umansky, 1976; Feldman, 1996, 1997; Pizlo et al., 1997; Smits & Vos, 1987) and Moiré patterns (Glass, 1969; Prazdny, 1984; Stevens, 1978). The computational literature has focused on contours (Guy & Medioni, 1996; Zucker, 1985) and surfaces (Barrow & Tenenbaum, 1981; Binford, 1981), whereas in the human vision literature, there is a widespread view that surfaces rather than objects are the primary unit of visual representation (Feldman, 1999). Throughout the above literature grouping is discussed independent of any temporal processes, just dependent on the spatial static features of the visual scene. On the other hand, objects have been the central focus in the developmental literature (Baillargeon, 1994; Spelke, 1990). There, interest has centered on defining properties of objects that go beyond strictly visual aspects, such as spatio-temporal cohesion and stability over time (e.g., Hock, Kelso & Schöner, 1993). Gibson (1979) discusses the salience of accretion and deletion of surfaces as important sources of visual information in spatiotemporal settings, such as moving observers.

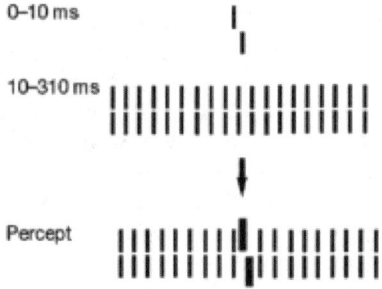

Figure 1. The 'shine-through' illusion is a spatiotemporal phenomenon. Initially, a first visual target is displayed very briefly for 10ms, followed by a second visual object for a longer period of 300ms. The resulting percept is plotted in the bottom row displaying a combination of the two preceding visual stimuli with enhanced features of the first stimulus. (Adapted after Herzog & Fahle 2002)

The perception of a visual object and the responses of cortical neurons can be strongly influenced by the context surrounding the object. In particular, in high-contrast regimes, the embedding of an object in an iso-oriented context reduces neural responses and deteriorates performance in psychophysical experiments. Performance based on orthogonal surrounds is better than from iso-oriented ones, a phenomenon postulated to be caused by long- or short-range interactions between neurons tuned to orientation (Nothdurft et al. 1999). Recently, Herzog & Fahle (2002) performed a spatio-temporal visual experiment and showed that the orientation difference between target and context does not determine performance. Instead, contextual modulation depends on the overall spatial structure of the stimulus. See Figure 1 for details. The authors make their case by establishing a phenomenon called the 'shine-through' effect, in which a briefly displayed target (duration 10ms) precedes a homogeneous grating appearing for a longer time period (duration 300ms). The target comprises two slightly shifted bars, which shine through in the succeeding grating and appear wider, brighter, longer and superimposed on the grating.

The authors discuss the emergence of this illusion in dependence of the spatial context only, but imply that the effect vanishes if the temporal features are altered appropriately. Max Wertheimer (1924), one of the founding fathers of Gestalt theory, says *"The basic thesis of gestalt theory might be formulated thus: there are contexts in which what is happening in the whole cannot be deduced from the characteristics of the separate pieces, but conversely; what happens to a part of the whole is, in clear-cut cases, determined by the laws of the inner structure of its whole."* Herzog & Fahle (2002) state this more specifically by postulating that the inner coherence of each, the target *and* the context, determines the effect of foreground-background separation that is the formation of the percept. Strictly speaking, the shine-through illusion is a spatiotemporal effect and sensitive to its timing properties such as duration and ISI.

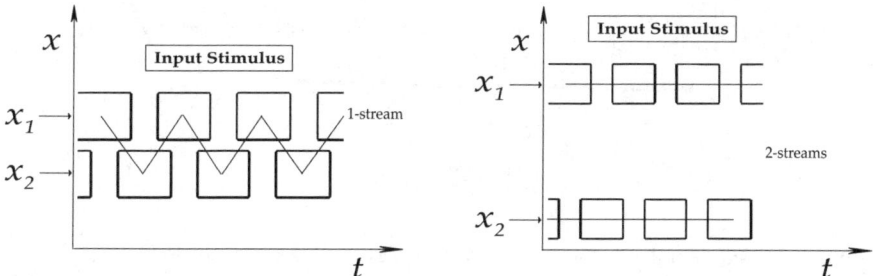

Figure 2. Streaming occurs in the auditory domain. Here, a sequence of tones of alternating pitch is presented to human subjects. On the left, the tone frequency is plotted vertically, and time is plotted horizontally. The sequence of tones is perceived as one stream of alternating tones as implied by the solid line connecting the stimuli. To the right, the same situation is shown, but for a greater pitch difference of the stimuli. Here, the percept changes and two independent sequences of tones of equal pitch are perceived, that is two streams.

Detailed studies of such timing effects have been performed extensively in the auditory domain (see e.g. Bregman, 1990), of which one of the most prominent is known as 'streaming'. In 'streaming' the percept depends on two control parameters, the ISI and the pitch difference of the two acoustic stimuli. For large ISIs and small pitch differences, the percept is a single stream of consecutive tones. In contrast, in the regime of short ISIs and sufficiently large pitch differences, in which the percept consists of two streams as two separate entities, the high-pitched tone and the low-pitched tone sequence. See Figure 2 for an illustration of the stimulus sequences.

Van Noorden (1975) identified three parameter regions (see Figure 3), in which either percept is stable exclusively (monostable) and a region in between, which is bistable and permits either percept, that is one-stream or two-streams.

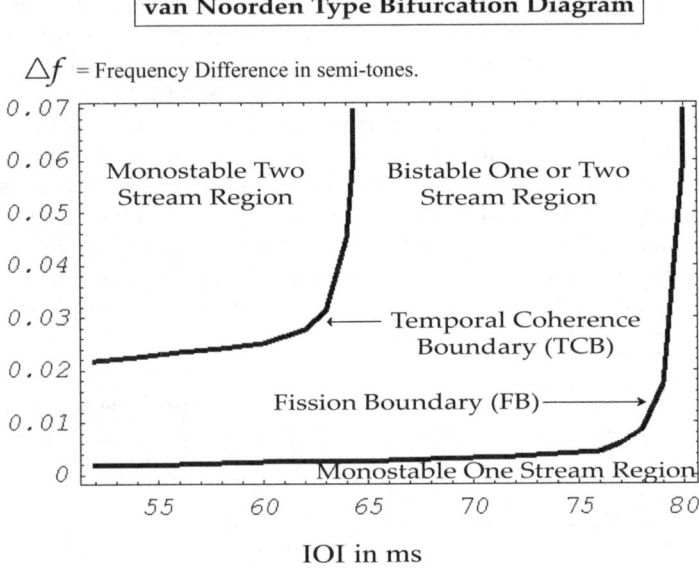

Figure 3. Van Noorden's parameter space displays two separate regions of monostable percepts that is either one or two streams only, with a region in between, which allows for bistable percepts.

In both of our examples, the 'shine-through' and 'streaming', manipulations have been performed in the temporal domain and in their individual spaces. These spaces are defined by the non-temporal qualities of the stimuli, such as spatial configuration, color or pitch. Expanding the idea of Herzog & Fahle (2002) of inner coherence to the temporal domain also, then the percept will change when the inner coherence along the temporal axis is lost while altering the control parameter in the temporal domain, here the ISI. This phenomenon has been observed in the example of 'streaming' when the distance of the stimuli along the temporal axis, the ISI, become too large (see Figure 2). We use the terminology of a distance deliberately, because in the following we are going to postulate a metric in an appropriate space, which shall provide a measure for the stability of percepts, or more generally, behavioral patterns which include perception and motor behavior. Recent studies by us (Jirsa et al., 2003) on inter-modal integration followed conceptually the lines of Van Noorden (1975) and identified parameter regimes in which the perception of auditory and visual stimuli changed qualitatively when varying the control parameters. Both modalities were presented periodically with the same ISI. However, the timing of the auditory and visual stimulus was varied such that they were shifted in time from dT=−200ms to dT=200ms with respect to each other.

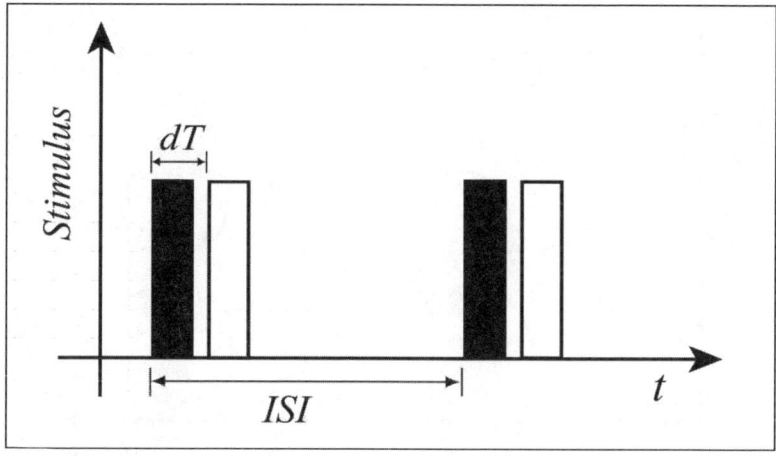

Figure 4. Multimodal stimulus sequences are presented which are comprised of an auditory stimulus (filled bars) and a visual stimulus (empty bars). Two variables characterize the sequence of stimulus, the time shift dT which is the intra-stimulus interval, and the interval between stimulus pairs, the inter-stimulus interval ISI.

Negative time shifts characterize a preceding auditory stimulus, an AV sequence, and positive time shifts characterize a preceding visual stimulus, a VA sequence. Two control parameters were varied, the time shift and the ISI (see Figure 4). In order to avoid the direction of attention towards either modality, the subjects were instructed 'to try to perceive the stimuli as simultaneous'. There will be conditions, in which this task cannot be done and a percept of simultaneity cannot be enforced. In such a situation, a spontaneous transition is made to a percept VA, AV or drift (D). This last percept, 'drift', is typically characterized by subjects either along the lines of a drift known from the relative phase in motor coordination, that is the stimuli appear to be alternating between phase locking and phase wrapping, or along the lines of streaming, that is the visual and auditory sequences appear as independent entities with uncorrelated phasings. Within the two-dimensional parameter space explored in this task, four regions with different percepts were manifest: an asymmetric region with the percept 'simultaneous' (S), flanked from the left and right by two regions with percepts AV and VA, respectively, and a region in which the percept of simultaneity was lost, but no particular order of the stimuli could be reported. This perceptual partitioning of the parameter space becomes an even more striking feature, if we substitute one modality, say the auditory stimulus, by another modality, say a tactile stimulus. First pilot data show strongest evidence that the partitioning is qualitatively independent of the modalities involved suggesting that these features are not modality specific, but rather a signature of the couplings involved (Jirsa et al., 2003).

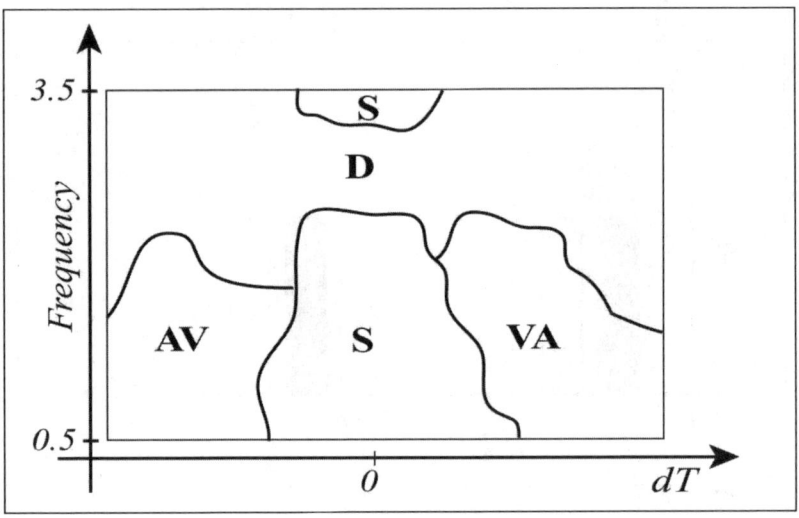

Figure 5. The intra-stimulus interval dT and the frequency (inverse of the inter-stimulus interval ISI) span a parameter space. The percepts obtained as a grand average from 16 subjects in a categorization task are segregated in four major sections with the percepts AV (auditory precedes visual stimulus), VA (visual precedes auditory stimulus), S (auditory stimulus coincides with visual stimulus) and D (there is no fixed order between the stimuli). Note the statistically significant area for large frequencies around dT=0 which is topologically separated from the low-frequency regimes.

If we recall the classic sensorimotor bimanual (e.g., Kelso, 1981, 1984; Kay et al., 1987) and unimanual synchronization-syncopation experiments (Kelso et al., 1990; see also Kelso et al., 2001), then the paradigm has always been to perform a unimanual or bimanual flexion-extension cycle either in-phase or anti-phase driven by a visual or auditory metronome beat. In dependence on external control parameters, such as the ISI, the stability of these coordination patterns (in- and anti-phase pattern) was altered. There has been much discussion on the nature of the processes causing stability changes, and thus qualitative changes, since the system performs a transition from one coordinated state to another. Perceptual (Mechsner, 2001, 2003), anatomical (Carson, 2000, 2003) and informational (Kelso 1995) effects have been proposed. These effects are not mutually exclusive and differ in importance in different realizations of experiments (for a nice demonstration, see Kelso et al., 2001). Qualitative changes in motor and perceptual behavior have been shown to be accompanied by simultaneous qualitative changes in the brain dynamics (Bushara et al., 2001; Daffertshofer et al., 2000; Fuchs et al., 1992; Jirsa et al., 1995; Jirsa et al., 1998; Kelso et al., 1991,

1992; Meyer-Lindenberg et al., 2002). Modeling approaches have predicted such qualitative changes in neural activity (Jirsa et al., 1994, 1998; Beek et al., 2002) postulating a causal role for the neural correlates, but not necessarily taking a stance on the nature of the underlying processes. The reason is that in the vicinity of such qualitative changes the dynamics of the neural activity has been shown to be governed by order parameters in the strict sense of Hermann Haken (1983, 1996). Order parameters may sometimes acquire more abstract and informational forms, of which the relative phase is only one example. Since qualitative change can be generically described within the framework of a bifurcation of an order parameter, this allows us to rephrase the issue of underlying processes in terms of the question: 'What makes a bifurcation bifurcate?' To address this question, the connection to the material substrate of the system must be made. It is here, where the discussion on the degree of the perceptual, anatomical and informational contributions may start. This is the ultimate goal of the hitherto cited papers on brain and behavioral dynamics. The other question arising is how the formation of a percept does affect the formation of a motor coordination pattern. In a recent paper, Bogaerts et al. (2003) studied the symmetry constraints in perception and action and concluded that the perceived geometry of objects correlates with the spontaneous selection of coordination modes. In this book chapter, we are less concerned with the constraining role of perceived spatial features on action, and more with the commonalities of the emergence of spatiotemporal patterns in the behavior in both, perception and action. Michael Turvey (2003) provides a beautiful exposition on how the perception-action divide dissolves within the dynamics and self-organization framework. Perception and action present themselves as components of one process, which is governed by the order parameter dynamics (Kelso et al., 1990; Turvey, 2003). Viewed from the perspective of coordination dynamics, we wish to identify the formation of a percept with the formation of a stable motor coordination pattern. Hence a percept is a stable coordination pattern itself, which may display all the dynamics known from the coordination of motor components such as transient dynamics, phase transitions, critical fluctuations and critical slowing down (see Kelso, 1995 for a review).

2. Grouping as a phase-locking phenomenon

From a theoretical perspective, the partitioning of the parameter space may actually be understood on the basis of synchronization and syncopation within a rhythmic paradigm. Yoshiki Kuramoto (1984) has shown that every set of two weakly coupled limit cycle oscillators displays synchronized and syncopated solutions, independent of the coupling. The coupling serves as a mechanism to control the stability of these solutions, but the solutions themselves always exist due to symmetry. Hence, imposing a rhythmic paradigm onto a behavioral experiment will always result in in-phase and anti-phase behavior. Its stability, however, will be a signature of the coupling. A classic example thereof is found in bimanual coordination, which displays bistability of in-phase and anti-phase solutions below a critical frequency and monostability of the in-phase solution above the critical frequency (see Swinnen, 2002 for a review). Haken, Kelso & Bunz (1985) discussed the dynamics along an explicit relative phase equation which reproduced the observed phenomena. The relative phase equation was derived from two oscillators, which were nonlinearly coupled by, what is nowadays known as the HKB coupling (Haken et al., 1985). Similar phase equations, together with many more details of the phase space trajectories, were obtained for parametrically excited systems (Jirsa et al., 2000), which model the coordination of one motor component with an external metronome (see also Kelso et al., 1990). Also, neurally-based modeling approaches provide similar results (see Nagashino & Kelso, 1993; Jirsa et al., 1998; Grossberg et al., 1997), and are just another expression of the phase equations' generics. Above we discussed the perceptual pattern formation under variation of the ISI and the time shift dT between two stimuli. In analogy, we wish to substitute one stimulus by a motor component and discuss the motor pattern formation under variation of the ISI and different initial time shifts dT. The latter serves as an initial condition, because when the initial coordination pattern set by dT is not stable, then the system will transient to the nearest stable stationary coordination state. As a consequence, the parameter space will also be partitioned in four sections as seen in Figure 6. Note that due to the periodic context the initial time shift dT is expressed in radians in Figure 6.

There are the two anti-phase regimes at the left and outer boundaries, the synchronization regime around dT=0ms and the phase wrapping or drift regime for greater stimulation rates. The parameter space in Figure 6 was generated using the relative phase equation for ϕ known from Haken et al. (1985) and extended by Kelso, et al (1990) to include the symmetry breaking term, $\Delta\omega$:

$$\dot{\phi} = \Delta\omega - \sin\phi - 2\sin 2\phi$$

The symmetry breaking term $\Delta\omega$ increases with the difference between the ISI and the characteristic frequency of the motor component, which is often identified with the most comfortable movement frequency. Hence the effects of increasing ISI and increasing $\Delta\omega$ are functionally equivalent for the scenario discussed in Figure 6.

Obviously, here we are attempting to identify commonalities between perceptual and motor pattern formation by equating the formation of a percept with the realization of a stable pattern and vice versa (see also Hock, Kelso & Schöner, 1993). The changes occurring along the vertical axis, the transition from phase

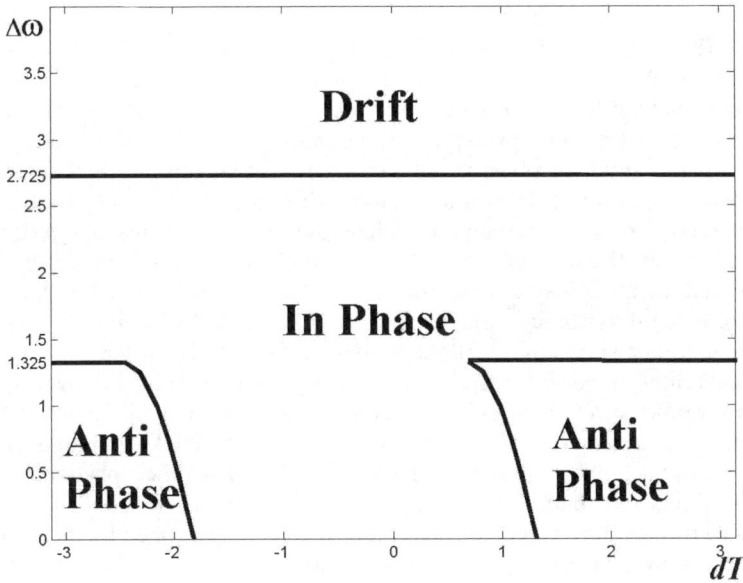

Figure 6. The stable stationary relative phase, obtained after a transient phase, is plotted in dependence of the initial conditions dT, that is the initial time shift between two oscillators, and a symmetry breaking term $\Delta\omega$, which characterizes differences between two oscillators, e.g. the difference between stimulus frequency and characteristic finger movement frequency.

locked solutions to phase drift, may only be understood on the basis of a symmetry breaking, which may have many realizations, such as non-identical oscillators or time-delay couplings. Experimentally, the degree of symmetry breaking and its effects have been systematically investigated (Carson et al., 2000, Lee et al., 2002) and captured theoretically by a generalized form of the HKB equations (Fuchs & Jirsa, 2001). The effect of symmetry breaking onto the dynamics is well known, that is a saddle-node bifurcation causes the loss of phase synchrony and the phases of the oscillators start drifting with respect to each other. (see Kelso, 1995, Chapter 4). Once the loss of phase locking has occurred, the individual phases show characteristic drifting behavior, in which stages of slow and fast phase drift alternate. Increasing the effect of symmetry breaking, the two oscillators lose also this last commonality of regimes, in which the oscillator phases move together for a finite time, and are completely segregated in two streams. The latter implication of streaming is intended. Both of these phenomena, drifting and streaming, have been reported by subjects in perceptual coordination experiments (Jirsa et al., 2003).

3. Grouping as a convergence phenomenon

When a rhythmic paradigm is not imposed onto a dynamic system, then a reduction to phase equations is not in general possible. In fact, almost all phenomena that we encounter in our environment on a daily basis, are of a discrete nature, whether they are perceptual or motoric in origin. But analogous to rhythmic paradigms, we observe the same types of integration and segregation phenomena. Naturally, the question arises whether the ideas developed in the previous section may be extended to include the dynamics of discrete phenomena as a special case. The answer is actually 'No', and the situation will rather have to be presented the other way around, that is the phase description is the special case of a more general dynamics. Gregor Schöner (1990) noted in a numerical study of two identical Gonzalez-Piro oscillators, coupled with an HKB term, that discrete movements either tend towards a parallel or a sequential behavior (Kelso, et al., 1979). In a more recent study (Jirsa & Kelso, 2003) we actually generalized this observation to arbitrary oscillators with sigmoidal and HKB-couplings (the so-called excitator). We showed mathematically that the phenomenon of parallelization/sequentialization may be understood as a convergence and divergence of trajectories in phase space and is solely determined by the properties of the coupling, independent of the intrinsic dynamics of the two identical oscillators.

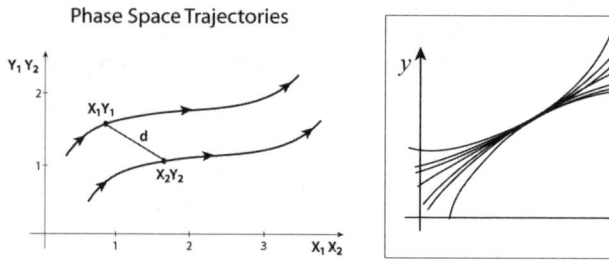

Figure 7. The time-dependent Euclidean distance d(t) in phase space provides a measure for the similarity of the dynamics of two systems given by the coordinates (x_1, y_1) and (x_2, y_2) as seen on the left. As these two systems evolve in time, convergent effects of their coupling tend to minimize the instantaneous distance between them and create trajectory bundles as seen on the right.

For discrete convergent dynamics, the couplings create something like a bottleneck in the phase space, in which all the trajectories are bundled. Mathematically speaking, the particular area in phase space serves as a manifold onto which the phase space dynamics reduces during a finite time. As a consequence, the distances, which are measurable by some reasonable metric such as the Euclidean distance d (see Figure 7 (left)), in the phase space become small as seen in Figure 7 (right).

The opposite is true for divergent phenomena, in which the couplings drive nearby trajectories away to maximize the distance. Periodic phenomena turn out to be a special case of convergence and divergence phenomena in phase space (see Figure 8): The system is forced upon a circular trajectory, the limit cycle, along which the shortest distance between two oscillators is for in-phase (convergence) and the longest distance is for anti-phase (divergence), that is half of the circumference. The same critical parameter values are obtained separating the convergent from divergent regime for both rhythmic and discrete dynamics.

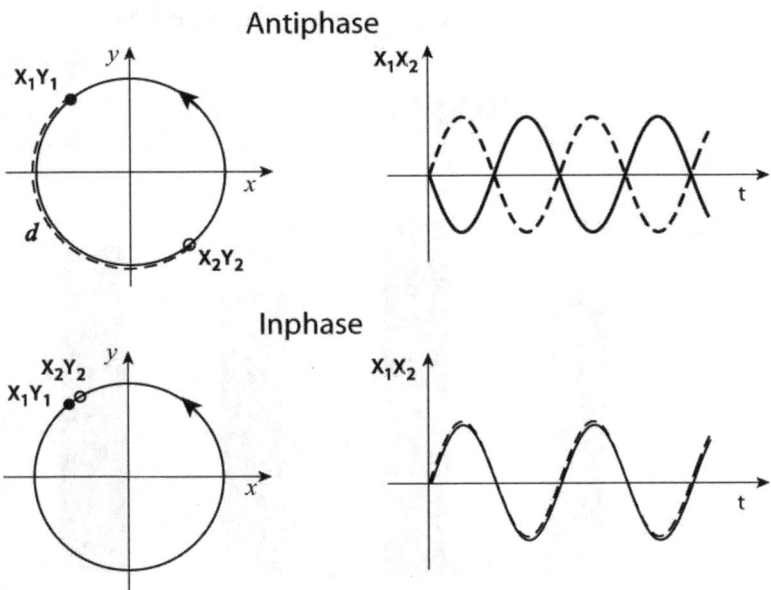

Figure 8. When rhythmic constraints are imposed on the experimental paradigm, then the two systems (x_1, y_1) and (x_2, y_2) become limit cycle oscillators. The phenomenon of divergence then results in anti-phase motion, because it maximizes the Euclidean distance $d(t)$ along the closed loop (top row). Equivalently, the phenomenon of convergence minimizes the Euclidean distance of two coupled oscillators and thus synchronizes their phases (bottom row).

4. Some thoughts on biological function

We have interpreted the phenomenon of spatiotemporal grouping as an integration of dynamic processes of the perceptual and motor kind towards a mutual dynamics, in which integration is to be understood as a convergence of trajectories in the phase space of the underlying dynamic system. Let us realize that such a convergence leads to a reduction of the number of degrees of freedom, since the system moves primarily, during the convergent phase, along a manifold, which means the system is constrained to a region of the phase space. Hence here we deal with a reduction of complexity, which may serve a beneficial function for the biological system. This is beautifully displayed in the following example. Human subjects show response times to auditory and visual stimuli typically on the order of 340ms and 370ms (stimulus on-set to peak movement), respectively. If each stimulus is augmented by another modality, such that a visual stimulus is accompanied by an auditory tone within a time window of 50ms and vice versa, then the response times are reduced up to 70ms as displayed in Figure 9.

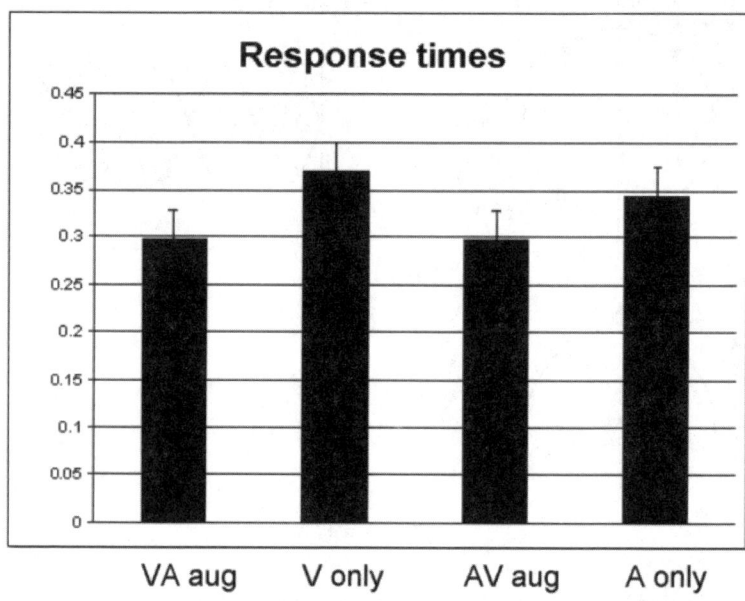

Figure 9. The mean response times (grand average over 12 subjects) to visual and auditory stimuli are shown, denoted by 'V only' and 'A only'. The augmentation of the relevant stimulus by a simultaneously occurring stimulus of the other modality reduces the response time significantly (shown as 'AV aug' and 'VA aug', respectively).

In our terminology we may interpret the reduced response time as the convergence of two stimulus trajectories and one movement cycle to a single integrated entity. On the other hand, if the two stimuli do not occur within a time interval of about 50ms, but with a greater inter-stimulus-interval such as 150ms, then the integration does not take place. On the contrary, segregation takes place and the response times are increased. This latter effect is known in the literature as the psychological refractory period (Davis, 1959). Such segregation is not necessarily disadvantageous for the individual, because it also frees processing resources. When two trajectories converge in phase space and move along a manifold, then they are not readily available for other tasks or behaviors. In fact, it would take a prolonged time to separate the two converged components before they can be used, or better integrated into other tasks. However, if these components are segregated from each other, then they are available immediately for other dynamic processes.

In this chapter we attempted to develop a line of thought, which unites perceptual and motor-behavioral phenomena that display grouping in space and time. Such a description is naturally amenable to the language of coordination dynamics. Many researchers have employed rhythmic paradigms in experimental tasks, as well as theoretical modeling, since it is well known that under these conditions the number of degrees of freedom is reduced. We show here that the language and the concepts developed within these rhythmic paradigms may be understood as a special case of convergent and divergent dynamic behavior. The more general situation may be the discrete behavioral dynamics. We wish to propose that the convergence and divergence of trajectories in phase space, such as the existence of areas with increased trajectory densities in phase space, provides a realization for a mechanism of integration and segregation.

Acknowledgments

We acknowledge the help of Collins Assisi during data collection and the help of Felix Almonte and Ajay Pillai for preparing some of the figures. Work supported by DARPA and NIMH.

References

Baillargeon R (1994) How do infants learn about the physical world? Curr Dir Psychol Sci 3 (5), 133-140

Barchilon Ben-Av M, Sagi D, Braun J (1992) Visual attention and perceptual grouping. Percept Psychophys 52 (3), 277-294

Barrow HG, Tenenbaum JM (1981) Interpreting line drawing as three-dimensional surfaces. Artif Intell 17, 75-116

Beek PJ, Peper CE, Daffertshofer A (2002) Modeling rhythmic interlimb coordination: Beyond the Haken-Kelso-Bunz model. Brain Cognition 48 (1), 149-165

Binford T (1981) Inferring surfaces from images. Artif Intell17, 205-244

Bogaerts H, Wagemans J, Meulenbroek RGJ, Van den Bergh O, Vangheluwe S, Puttemans V, Wenderoth N, Swinnen SP (2003) Bimanual production of triangular drawing patterns: Exploring symmetry constraints in perception and action. J Motor Behav (in press)

Bregman AS (1990) Auditory Scene Analysis. MIT Press, Cambridge Massachusetts

Bushara KO, Hanakawa T, Immisch I, Toma K, Kansaku K, Hallett M (2002) Neural correlates of cross-modal binding. Nat Neurosci 6 (2), 190-195

Caelli TM, Umansky J (1976) Interpolation in the visual system. Vision Res16 (10), 1055-1060

Carson RG, Riek S, Smethurst CJ, Lison JF, Byblow WD (2000) Neuromuscular-skeletal constraints upon the dynamics of unimanual and bimanual coordination. Exp Brain Res 131, 196-214

Carson RG (2003) Governing coordination. Why do muscles matter? This volume

Daffertshofer A, Peper CE, Beek PJ (2000) Spectral analyses of event-related encephalographic signals. Physics Letters A 266 (4-6), 290-302

Davis R (1959) The role of "attention" in the psychological refractory period. Q J Exp Psychol 11 (4), 211-220

Feldman J (1996) Regularity vs Genericity in the perception of collinearity. Perception 25, 335-342

Feldman J (1997) Curvilinearity, covariance, and regularity in perceptual groups. Vision Res 37 (20), 2835-2848

Feldman J (1999) The role of objects in perceptual grouping. Acta Psychol 102, 137-163

Fuchs A, Kelso JAS, Haken H (1992) Phase Transitions in the Human Brain: Spatial Mode Dynamics. Int J Bifurcat Chaos 2, 917-939

Fuchs A, Jirsa VK (2000) The HKB Model revisited: How varying the degree of symmetry controls dynamics. Hum Movement Sci 19, 425-449

Gibson JJ (1979) The ecological approach to visual perception. Houghton Mifflin, Boston

Gilchrist AL (1977) Perceived lightness depends on perceived spatial arrangement. Science 195, 185-187

Glass L (1969) Moiré effects from random dots. Nature 223, 578-580

Gonzalez DL, Piro O (1987) Global bifurcations and phase portrait of an analytically solvable nonlinear oscillator: Relaxation oscillations and saddle-node collisions. Phys Rev A 36, 4402-4410

Grossberg S, Pribe C, Cohen MA (1997) Neural control of interlimb oscillations I. Human bimanual coordination. Biol Cybern 77, 131-140

Guy G, Medioni G (1996) Inferring global perceptual contours from local features. Int J Comput Vision 20, 113-133

Haken H (1983) Synergetics. An Introduction. 3rd edn. Springer, Berlin Heidelberg New York

Haken H (1996) Principles of brain functioning. Springer, Berlin Heidelberg New York

Haken H, Kelso JAS, Bunz H (1985) A Theoretical Model of Phase transitions in Human Hand Movements. Biol Cybern 51, 347 – 356

Herzog HH, Fahle M (2002) Effects of grouping in contextual modulation. Nature 415, 433-436

Hock HS, Kelso JAS, Schöner G (1993) Bistability, hysteresis, and phase transitions in the perceptual organization of apparent motion. J Exp Psych Human 19, 63-80

Jirsa VK, Friedrich R, Haken H, Kelso JAS (1994) A theoretical model of phase transitions in the human brain. Biol Cybern 71, 27-35

Jirsa VK, Friedrich R, Haken H (1995) Reconstruction of the spatio-temporal dynamics of a human magnetoencephalogram. Physica D 89, 100-122

Jirsa VK, Fuchs A, Kelso JAS (1998) Connecting cortical and behavioral dynamics: Bimanual coordination. Neural Comput 10, 2019-2045

Jirsa VK, Fink P, Foo P, Kelso JAS (2000) Parametric stabilization of biological coordination: A theoretical model. J Biol Phys 26, 85-112

Jirsa VK, Assisi CG, Dhamala M, Kelso JAS (2003) unpublished data

Jirsa VK, Kelso JAS (2003) The Excitator as a minimal model for discrete and rhythmic movement generation. Submitted

Kay BA, Kelso JAS, Saltzmann EL, Schöner G (1987) Space-time behavior of single and bimanual rhythmical movements: Data and limit cycle model. J Exp Psychol 13, 178-192

Kelso JAS (1981) On the oscillatory basis of movement. B Psychonomic Soc 18, 63

Kelso JAS (1984) Phase transitions and critical behavior in human bimanual coordination. Am J Physiol 15, R1000-R1004

Kelso JAS, Southard D, Goodman D (1979) On the nature of human interlimb coordination. Science 203, 1029-1031

Kelso JAS, DelColle JD, Schöner G (1990) Action-perception as a pattern formation process. In: Jeannerod M (ed.) Attention and performance XIII. Erlbaum, Hillsdale, NJ, 136-169

Kelso JAS, Bressler SL, Buchanan S, DeGuzman GC, Ding M, Fuchs A., Holroyd T (1991) Cooperative and critical phenomena in the human brain revealed by multiple SQUIDS. In: Duke D, Pritchard W (eds.), Measuring Chaos in the Human Brain. World Scientific, New Jersey, 97-112

Kelso JAS, Bressler SL, Buchanan S, DeGuzman GC, Ding M, Fuchs A, Holroyd T (1992) A phase transition in human brain and behavior. Phys Lett A 169, 134-144

Kelso JAS, Fink P, DeLaplain CR, Carson RG (2001). Haptic information stabilizes and destabilizes coordination dynamics. P Roy Soc B Bio 268, 1207-1213

Kelso JAS (1995) Dynamic Patterns. The Self-Organization of Brain and Behavior. The MIT Press, Cambridge, Massachusetts

Kubovy M, Holcombe AO, Wagemans J (1998) On the lawfulness of grouping by proximity. Cognitive Psychol 35, 71-98

Kuramoto Y (1984) Chemical oscillations, waves, and turbulence. Springer Berlin Heidelberg New York

Lee DT, Quincy JA, Chua R (2002) Spatial constraints in bimanual coordination: influences of effector orientation. Exp Brain Res 146, 205-212

Mayville JM, Jantzen KJ, Fuchs A, Steinberg FL, Kelso JAS (2002) Cortical and subcortical networks underlying syncopated and synchronized coordination revealed using fMRI. Hum Brain Mapp 17, 214-229

Mechsner F, Kerzel D, Knoblich G, Prinz W (2001) Perceptual basis of bimanual coordination. Nature 414, 69-73

Mechsner F (2003) A perceptual-cognitive approach to bimanual coordination. This volume

Meyer-Lindenberg A, Ziemann U, Hajak G, Cohen L, Berman KF (2002) Transitions between dynamical states of differing stability in the human brain. Proc Natl Acad Sci USA 99, 10948-10953

Nagashino H, Kelso JAS (1992) Phase transitions in oscillatory neural networks. Science of Artificial Neural Networks. SPIE 1710, 278-297

Nothdurft HC, Gallant JL, van Essen DC (1999) Response modulation by texture surround in primate area V1: Correlates of pop-out under anesthesia. Visual Neurosci. 16, 15-34

Pizlo Z, Salach-Golyska M, Rosenfeld A (1997) Curve detection in a noisy image. Vision Res 37 (9), 1217-1241

Prazdny K (1984) On the perception of Glass patterns. Perception 13, 469-478

Schmidt RC, Shaw BK, Turvey MT (1993) Coupling dynamics in interlimb coordination. J Exp Psychol Human 19, 397–415

Schöner G (1990) A dynamic theory of coordination of discrete movement. Biol Cybern 63, 257-270

Smits JT, Vos PG (1987) The perception of continuous curves in dot stimuli. Perception 16, 121-131

Spelke ES (1990) Principles of object perception. Cognitive Sci 14, 29-56

Sternad D, Dean WJ, Schaal S (2000) Interaction of rhythmic and discrete pattern generators in single-joint movements. Hum Movement Sci 19, 627-664

Stein BE, Meredith MA (1993) The Merging of the Senses. MIT Press, Cambridge, Massachusetts

Stevens KA (1978) Computation of locally parallel structure. Biol Cybern 29, 19-28

Swinnen SP (2002) Intermanual Coordination: From behavioural principles to neural-network interactions. Nat Rev Neurosci 3, 350-361

Turvey MT (2003) Impredicativity, Dynamics, and the Perception-Action Divide. This volume

Van Noorden LPAS (1975) Temporal Coherence in the Perception of Tone Sequences. Unpublished doctoral dissertation, Eindhoven University of Technology

Wertheimer M (1924) Gestalt theory. Social Research 11 (translation of lecture at the Kant Society, Berlin, 1924)

Zucker SW (1985) Early orientation selection: Tangent fields and the dimensionality of their support. Comput Vision Graph 32, 74-103

Zucker SW, Stevens KA, Sander P (1983), The relation between proximity and brightness similarity in dot patterns. Percept Psychophys 34, 513-522

Author Index

A
Adamovich SA 171
Albert R 202
Alkire MT 236
Allum JH 104
Almeida QJ 47
Amazeen EL 24, 43-44, 47, 183
Arutyunyan RH 113
Asatryan DG 158, 160

B
Baillargeon R 244
Balasubramaniam R 155, 157, 158, 169
Baldissera F 148
Barchilon Ben-Av M 244
Bardy BG 103-109, 114-115
Barnes R 231-232
Barrow HG 244
Barth DS 236
Barto AG 93
Beek PJ 12, 107, 116, 183, 249
Bernstein NA 46, 49, 52, 57, 84, 113, 156, 159
Berthenthal BI 92
Bhushan N 156
Binford T 244
Blinkenberg M 143
Bogaerts H 243, 249
Bootsma RJ 92, 97, 104, 106, 114, 116
Bose A 235
Brashers-Krug T 34, 58
Bregman AS 244-245
Bressler SL 197-198, 201, 211, 218
Buchanan JJ 104, 107, 183
Bushara KO 248
Byblow WD 42-43, 47, 148-149, 151, 187

C
Cabrera JL 99
Caelli TM 244
Carroll TJ 144
Carson RG 24, 43, 46, 48-49, 69, 103, 142-145, 147-150, 187, 189-192, 248, 251
Cassirer E 15
Cattaert D 183
Chardenon A 37
Cheney PD 142
Cohen L 179-180
Corna S 104
Cottingham J 6
Crick F 233, 239
Crovitz F 222, 239

D
Daffertshofer A 248
Dai TH 143
Darwin C 178
Davis R 243, 254
deCharms RC 198
Descartes R 1, 3, 6
Dettmers C 143
Dewey J 10
Diedrich FJ 31, 103
Dijkstra TMH 91, 103, 111, 124-125
Ding M 198
Dukas R 229-230, 232

E
Eck D 223
Eddington A 15-17
Ehrlacher C 115
Eigen M 129, 177
Ettema GJC 148

F

Fagard J 47
Faugloire E 114
Feige B 143
Feldman AG 156, 158, 160, 166, 167, 169
Feldman J 243-244
Felleman DJ 197, 2002
Fitch H 10
Fitts PM 187
FitzHugh RA 219-211, 227-228, 235
Fitzpatrick RC 114, 183
Fontaine RJ 48, 63, 113
Foo P 93, 94, 96-100
Fourcade P 113, 116
Fowler CA 10, 171
Fox JJ 235
Frackowiak RSJ 197
Friston KJ 198, 201, 211
Fuchs A 35, 66, 248, 251

G

Gelfand IM 160
Ghafouri M 173
Gibson JJ 11, 244
Gilchrist AL 243
Glansdorff P 158
Glass L 244
Gonzalez DL 252
Gordon AM 144
Gray CM 198, 222, 233
Greene LS 47, 230
Greenwood PE 230
Grossberg S 287, 250

H

Haken H 1, 13, 22, 23, 24, 59, 60, 63, 103, 105, 180, 249-250
Hanson NR 9
Herzog HH 244-246
Hilgetag CC 202
Hobbes T 16

Hock HS 244, 250
Hodges NJ 50, 51, 63
Hodgkin AL 219-220, 235
Hollerbach JM 156
Hommel B 187, 192
Horak FB 104, 107, 124
Hoyt DF 31
Humphrey DR 143

J

James W 49, 229
Jantzen KJ 35-36, 188
Jeka JJ 66, 92, 103, 123-124, 126, 148
Jirsa VK 66, 197, 203, 211, 243, 246, 248-252
John ER 236
Jones DS 204
Jones MR 229-232

K

Kahneman D 25
Kawato M 156
Kay BA 248
Keijzer F 178
Kelso JAS 1, 5, 13, 14, 17, 18, 21-24, 28, 30, 31, 33, 35, 41-43, 48, 49, 59-63, 66, 68, 69, 73, 74, 91, 93, 100, 103, 104, 107, 113, 114, 146-149, 174, 178-180, 182, 183, 191, 197-199, 201, 203, 211, 217
Kinsbourne M 143
Kline M 7
Kötter R 198
Kubovy M 244
Kugler PN 11, 59
Kuramoto Y 249
Kurtz S 48
Kwakernaak H 92

L

Lamme VAF 218, 234

Large EW 221, 229-230, 232
Lashley KS 156, 171
Latora V 212
Lee DN 111
Lee TD 24, 41, 43, 48-51, 63, 84, 92, 99, 110, 149, 182, 251
Lemon R 143
Lestienne F 111, 164-166
Levin MF 157-158, 163-164
Lippa Y 187
Lisman JE 224-225
Lockhead GR 227
Longtin A 220-221, 235
Lumer ED 235-236

M
March A 16
Marin L 104, 106, 114
Matthews PBC 158
Mayr E 178
Mayville JM 35
McCollum G 104-105, 107, 109
McDonald KD 136
McDonald PV 113
McIntosh AR 197-198
Mechsner F 4, 141, 146-147, 150, 177, 181, 183, 185-186, 189, 199, 248
Merikle PM 222
Mesulam MM 197
Metzger W 192
Meyer-Lindenberg A 249
Miller GA 41-52
Miltner WHR 234
Mittelstaedt H 156-157
Monno A 31, 34, 37, 44, 79
Mumford D 237

N
Nagashino H 250
Nagumo J 220-221, 227-228, 235
Nashner LM 104-105, 107, 109, 123
Navon D 25

Newell KM 57, 105-106, 113, 125
Nothdurft HC 244
Nunez PL 223-224, 227, 234

O
Ogata K 98
Ostry DJ 156
Oullier O 111, 112, 116, 188, 189

P
Païï Y-C 104
Palmer E 58, 142
Parasuraman R 25
Pastis OM 13, 51
Pellionisz A 3
Peters M 43
Pigeon P 168
Pizlo Z 244
Powers WT 178
Prazdny K 244
Prinz W 4, 49-50, 181

R
Rees G 238
Riccio GE 106, 110
Riek S 142
Rock I 192
Rodriguez E 239
Roelfsema PR 198
Rosen R 1, 5, 6, 8, 9, 10, 16-18
Rosenbaum DA 110
Rossi E 171, 173
Russell B 7, 81, 16

S
Saltzman EL 104, 171
Sanders AF 25
Savelsbergh GJP 92, 97
Scannell JW 197, 202
Schieber MH 143
Schlaug G 143

Schmidt RC 28, 84, 103, 107, 183
Scholz JP 22, 24, 28, 42, 168, 173, 182
Schöner G 1, 5, 13-14, 22-25, 28, 33, 59-61, 66, 74, 92, 103, 107, 11, 113, 123-124, 128, 168, 244, 250, 252,
Semjen A 150, 183
Serrien DJ 47
Shadmehr R 34, 58, 156
Shaw RE 10, 59, 66
Sherman SM 236
Simon HA 224
Simon JR 187
Singer W 198, 222, 233
Smethurst CJ 43, 46, 48-49, 69
Smits JT 244
Spelke ES 244
Sperling G 229, 232
Sporns O 197, 200-202, 210-212
Srinivasan R 198, 234
Stein BE 198, 243
Sternad D 107
Sternberg S 222, 224, 238
Stevens KA 244
Stevens SS 226-227
Stoffregen TA 106, 110, 114, 116
Strogatz SH 202
Summers JJ 150
Swinnen SP 24, 42, 47-49, 148, 183, 250

T

Temprado JJ 21, 26, 28, 29, 33, 36, 44-45, 47, 79
Tolat VV 160
Tononi G 197-203, 212, 234-236
Tsang PS 25
Turvey MT 1, 10-11, 42, 59, 66, 79, 183, 249

V

Vallbo ÅB 142
van Asten WNJC 11

van der Molen MW 25
van der Pol B 218-219, 230
van Noorden LPAS 246
Varela F 197-199, 234
Vereijken B 113
Verschueren SMP 47-48, 113
von Holst E 156-157, 160
von Stein A 198

W

Wagemans J 184, 244
Ward LB 58
Ward LM 198, 217, 226-233, 236-237
Warren WH 12, 31, 97, 103
Watts DJ 202
Weeks DJ 186-187, 190
Wegner D 239
Wenderoth N 48-49, 51, 63, 113
Wertheimer M 295
West RL 226, 229
Wilson HR 204
Wilson K 13
Wimmers RH 183
Wishart LR 43, 47, 49
Wolpert DM 156
Won J 171
Woollacott MH 104, 123
Wulf G 49-50
Wuyts IJ 22-24

Y

Yamanishi J 47, 150
Yoneda S 104
Young MP 197, 202
Yue GH 143

Z

Zanone PG 22, 24, 28, 33, 37, 44, 48-49, 57, 59, 61-63, 68-69, 73-75, 79, 82, 91, 113-114
Zeki S 237
Zucker SW 244

Index

A
Action 1-3, 38, 47, 59, 66, 87-88, 96-97, 101, 109, 118, 120-121, 137-139, 141, 143, 147, 151, 156-157, 160, 174, 177-178, 192, 203, 223, 227, 236, 239, 249, 255, 257, 259
Afferent
   input 137, 161-162, 170
   signal 156, 162
Aging 53, 55, 123, 139
Attention 20-22, 24-27, 43-46, 79-80, 82, 86-87, 190, 218, 229-232, 239, 242, 247, 255
   directed 43, 47, 52
   focus 31
Attentional
   cost 21-22, 25-28, 79-82
   entrainment 229
   focus 31-32, 41, 232
   load 21-22, 30, 32, 39, 55, 88
   resources 25, 36, 80 ,229
Attractors 12 ,14, 23, 39, 55-56, 61-64, 71, 101, 111, 121
   basin of attraction 61, 72-74, 77, 79, 84

B
Balancing 50, 91, 93
Bell-Magendie   law 2
Bifurcation 60, 74, 76, 84, 116, 249
Bifurcation
   saddle node 251, 256
Binding problem   211, 233, 244, 256
Biological 6, 8, 14, 38, 60, 104, 136, 160, 174, 177-178, 193-194, 239, 253
   coordination 2, 20, 153, 257

systems 20, 21, 116, 158, 175, 178, 254
Biomechanics 14
Brain imaging 36, 239
   EEG 35, 223-224, 234, 238, 241
   fMRI 35, 143, 258
   MEG (SQUID) 234, 238

C
Cartesian machine 5, 6, 8, 15
Causal sequences 17
Causality 6, 9
   circular 22, 113
Center of mass   106-107, 119, 128, 130, 132
Central nervous system (CNS)   19, 21, 25, 28, 30, 109, 142, 148, 151, 175
Central pattern generator 157
Cerebrum 2
Cognition 16, 21-22, 177, 193, 198, 213-215, 217, 242
Coherence 83, 198, 213, 234, 241, 245-246, 259

Command 4, 163, 179
   central 104, 171
   coactivation 163-164
   motor 144, 183, 189
   mucular 124
Competition regime 74
Components 4, 6, 14-15, 21-23, 28, 46, 59, 60, 63, 107, 110, 118, 124-125, 129, 139, 155-156, 160, 171, 200, 243, 249, 255
Computation   3, 18, 99, 132-133, 155-156, 173-174, 210-211, 221, 241, 244, 259
Connectivity 201

Conscious awareness 218, 222, 234
Consciousness 46, 52, 217, 222 - 223, 234, 236, 239
Consolidation 34, 37-38, 58, 82-83, 85
Constraints 21, 24, 37, 54, 59, 104-105, 123, 141, 170, 178, 184
   environment 106
   interaction 108
   motoric 183
   musculo-skeletal 141-147, 152
   perceptual 183
   physical 192
   rhythmic 253
   symmetry 249
   task 49, 161, 163
Context dependence 6-8, 13-14, 141-142, 144, 146-149, 151, 193, 234 , 244-245, 257
Control 6, 4, 14, 39, 91, 98, 103, 107-108, 124, 177
   level 155, 159, 177
   parameter 23, 42, 155, 244-245
Convergence 16, 61, 235, 244, 252-253
Cooperation regime 74
Coordination 10, 41, 57, 92, 141, 174, 201
   interpersonal 183, 189
   bimanual 21-22, 41, 43, 62, 145, 178, 183, 250
   biological 2, 20
   discrete 252
   dynamics 12-13, 47, 59, 65, 103-104, 174, 217, 243, 249
   interjoint 171
   interlimb 21, 28, 103, 147
   intralimb 21, 28, 103, 183
   learning 41
   mirror mode 177-185, 190
   modes 104, 107, 142, 185, 249
   motor 21, 247, 249
   parallel mode 146-148, 171, 179-183, 185, 190
   patterns 21, 23, 42, 68, 142, 144, 248
   perceptual 187, 251
   polyrhythm 4, 86
   postural 103-106
   rhythmic 183
   sensorimotor 141
   states 21
Correlation 97, 100, 198, 204, 208-209, 234
Coupling 20, 22-23, 94, 106, 235, 248-249
   global 200
   HKB 250-252
   informational 28
   mechanical 94
   neural 147-149
   perception-action 96-101, 192-193
   postural 111
   time-delay 251
   transient 201
   visual 110-112
Critical slowing down 106, 109, 249

D

Decay 129-130, 132, 219
Degrees of freedom 10, 20, 103, 113, 120, 153, 156, 168, 254-255
Delay 24, 31, 36, 43, 83, 97, 114, 129-130, 235, 240, 251
Descartes 1, 3, 6, 18
Divergence 243, 252-253
Dukas and Clark model 229, 232
Dynamic grouping 243
Dynamical systems approach 21, 36, 193
Dynamics 2, 12, 21, 43, 58, 125, 133, 148, 152, 178, 187, 200, 249
   anticipatory 99
   brain 201, 211, 248
   coordination 12-13, 21-22, 25, 27, 56, 59, 101, 103, 174, 180, 215, 217, 243
   discrete 253

intrinsic 47, 52, 64, 74, 103, 114, 252
learning 59
linear 203
multistable 68, 73
neural 197, 212
nonlinear 11, 116, 213
transient 249
translational 98

E
Ecological psychology 59
Electromyographic signals (EMG) 156, 158
Emergence 14, 37, 57, 60, 104-105, 198-199, 202, 243, 245, 249
Entailment 6-7, 9-10, 14
  impoverished 17-18
Entraining sequence 231-232
Environment 3-4, 8, 12, 22, 49, 58-59, 66, 106-107, 112, 124, 156, 160, 163, 232, 252
Epistemology 1, 19
Equilibrium point hypothesis 158
Equilibrium state 156, 160, 171
Evoked potential 145
Excitator model 257
Excitatory activity 207
Excited systems 250

F
Feedback 49, 114, 141, 234
  afferent 161
  augmented 47-48, 51, 149
  movement 157
  proprioceptive 157, 161, 175
  signal 48
  terminal 48
Feedforward processes 159, 234
FitzHugh-Nagumo model 220-221, 227-228, 235, 241
Fluctuations 23, 26, 29, 64, 175, 230, 232

Fluctuations critical 108-109, 116, 194, 249
Force gravitational 22, 64, 170
Foreaging behavior 229
Foreground-background separation 245
Forgetting 60, 66, 75
Frequency 23, 106, 248
  critical 46, 250
  eigen frequency 107, 130
  gamma 224, 233-234
  locked 4
  oscillation 22-23, 28-29, 62, 64, 114, 182
Function
  interaction 210-211
  natural 111
  sensorimotor 198
  somatosensory 125
  theta 224

G
Generic principles 58, 83
Gestalt 86, 192, 194, 198, 233, 245, 259
Gestalt Theory 194, 245, 259
Goal 41, 43, 46-47, 93, 110, 113, 119, 156, 160, 178, 183, 232
Goal-directedness 53
Gödel's Theorem 8
Ghost in the machine 1, 3

H
Handedness 37, 43, 47, 53-54
Haptic 54, 92-93, 138, 149, 153, 258
Hereditarily 6
HKB model 22-24, 36, 63, 71, 73, 85, 250-252, 256
Hodgkin-Huxley model 219, 223, 230, 242
Hysteresis 108-109, 116, 257

I

Impredicativity 1, 259
Information 11-12, 19, 21, 28, 36, 43, 47, 75, 92, 100, 125, 178, 198, 212-213
   behavioral 24, 28, 33, 47, 113, 218
   haptic 91, 93, 194, 258
   multisensory 123, 132
   mutual 200
   perceptual 59, 97-98, 100
   sensory 123, 127-128, 237, 244
   somatosensory 118
   velocity 92
   vestibular 125, 127
Instant equilibrium point (IEP) 125
Instructions 4, 38, 41-43, 49-51, 53-54, 74-75, 81, 86, 181, 187, 194
Integration 197-213, 243, 252
   inter-modal 246
   multisensory 139
   perceptual 253
Intention 4, 11, 21-22, 24-25, 41-43, 46, 59, 103, 105, 110-112, 116, 156, 178
Interceptive action 97
Inter-stimulus interval 243, 248, 254
Invariant trajectory 168

J

Joint redundancy 155, 158, 170

K

Kinematics 119, 128, 155-156, 158, 160, 173
Knowledge of results (KR) 34, 37, 68, 71, 75, 80-81

L

Lambda model 155-163
Law Bell-Magendie 2
Learning 24, 33-34, 36-37, 41, 47, 58-59, 100-101, 105, 113-115, 156, 173, 177, 179, 234
   route to 76-77, 80, 82, 84
   style 49
   transfer of 39, 50-51, 68-84, 119, 121
Linear causal chain 2
Lissajous figure 48, 50
Load
   attentional 21, 25, 31, 32, 34, 55, 88
   mental 79
Loops 8-9, 178, 222, 234, 237, 241
   closed 6-7, 19, 125, 137, 253
   open 125, 137

M

Manifold 155, 168, 173, 174, 184, 192, 201, 252, 254
   uncontrolled 155, 173, 176
Memory 22, 57-60, 65, 119, 156
   reorganization 59
   short term 217, 224-225, 238
   variable 65-67, 73
Metastability 197-199, 201-213, 235
Metric tensor 3, 4
Model
   lambda 155-163
   Dukas and Clark 229, 232
   excitator 257
   FitzHugh-Nagumo 220-221, 227-228, 235
   HKB 22-24, 36, 63, 71, 73, 85, 250-252, 256
   Hodgkin-Huxley 219, 223, 230, 242
   noisy computational 132, 135-136
   predator vigilance 229-230
   van der Pol 218-219, 223, 230, 242
Modes 107, 116
   anti-phase 104-105

in-phase 104-105, 114
coordination 142
spatial 256
Motor 2, 3, 10, 198, 201, 234, 253
  behavior 43, 173, 243, 246
  goal 156
  language 3
  memory 156
  neurons 142, 144, 153, 155, 159, 161-162, 170
  progams 84, 86, 110
  units 142-144, 159, 173
Movement
  continuous 13
  discrete 79, 80, 232, 243, 252
  multistability 14, 68, 73, 82
Muscles
  antagonist 22, 62, 147, 159, 161, 163-164, 170, 175, 199
  homologous 22, 26, 41, 46, 62, 146, 180, 182, 184, 188-189, 193-194
  synergies 147, 163, 169

N

Nervous system 19, 21, 22, 99, 109, 125, 136, 141, 148, 156-157, 160, 162-164, 173-175, 213, 215
Neural networks 55, 100-101, 154, 176, 195, 215, 223, 235, 258-259
Neurons 100, 137, 142, 144, 152-153, 159, 161-163, 170, 179-180, 197-198, 201-206, 210, 211, 213, 214, 219, 221-227, 233-241, 244
Newtonian 16, 18
Noise 23, 64, 99, 124, 125-127, 129-139, 203, 204, 209, 220, 228, 230, 232-233
  1/f 233
  process 133, 136
  pink 233
  white 64, 133
Non linear dynamics 201, 203, 213

O

Optic flow 11, 91
Oscillations
  alpha 37
  attentional 229-230
  beta 37, 143
  damped 128-136
  gamma 223-224, 233
  global 222-223
  harmonic 85
  local 223-224
  relaxation 217-239, 256
  van der Pol relaxation 219, 239, 2424
Overdamped motion 22

P

Paradigm 21, 25, 29, 31, 36, 69, 79, 82, 91, 104, 11, 116, 120, 139-142, 149-150, 179, 184, 187, 189, 192, 219, 243, 248-249, 252-253, 255
Parameter
  adjustment 99
  control 23, 42, 108, 156, 178, 244-248
  order 13, 118, 193, 249
Patterns
  adaptive 111
  anti-phase 22-35, 42-51, 62-63, 68, 71, 80, 111, 248
  behavioral 13-14, 33, 57, 59, 62-64, 86, 178, 246
  complex 197
  in-phase 42-43, 48, 62, 68, 71, 80, 104-105, 107-108, 111, 114, 248-250, 253
  neural 203
Pendulum 5, 8, 50, 11, 116, 133, 136, 183, 218
Percept formation 244
Perception 1-20, 38-39, 57, 59, 87-88, 96, 108, 116, 118, 120-121, 137-139, 141, 151, 174, 177, 183,

188, 192-194, 197-198, 215, 241-243, 246, 249, 255-259
perceptual coordination 187, 194, 251
perceptual grouping 184, 215, 243, 255, 256
perceptual variables 95, 97, 98
Perception-action 1-19, 1018, 152, 192-193, 249, 259
divide 1-19, 249, 259
Perceptual-cognitive 177-195
Phase
drift 47, 93-98, 247, 251
driving 248
locking 105, 220, 241-247, 249, 251
relative phase 22, 23, 26, 28-35, 42, 48, 50, 54, 62-63, 67-71, 75-76, 86, 104, 107-116, 149-150, 153, 181-186, 230-231, 247, 249, 251
transition 31-34, 37-39, 49, 54, 56, 60, 80, 87, 89, 101, 103, 114, 117-118, 120-121, 152-153, 183, 193-195, 249, 256-258
wrapping 247, 250
Posture 14, 91, 100, 103-104, 110-114, 116-121, 123-124, 137-139, 142, 145, 155-160, 174
postural coordination 103-121
intention 105, 110-112, 116
learning 105, 113-117
movement problem 156-158, 160
self-organization 105-107, 11, 113, 116
sway 101, 110-11, 121, 123-132, 136-139
vestibular loss 118, 124
Potential 22, 24, 28
Power spectrum 1/f 232-233
Predator vigilance model 229, 230
Primitives 155, 173
Principle 4, 7-8, 14, 16, 20-22, 36, 54-55, 57-58, 61, 63, 65-66, 69, 71, 78, 89, 103-104, 113, 116, 154-162, 171, 176-181, 184, 188-

189, 192-195, 197-198, 202, 211, 240, 257-259
coordination 22
minimization 163
Process
consolidation 34, 37-38, 58, 82-83, 85, 88
fusing 123, 125, 127, 129, 131-133, 135-139
linear 127-128
reorganization 38, 58-59, 114
stochastic 126-127, 138
Programming 157, 160
Proprioception 183
Psychological refractory period 254
Psychological scaling 225
Psychophysics 226, 242
Pursuit tracking task 48, 103-119

R

Reaction time 21, 25, 29, 32-36, 80, 187, 230, 233, 242-243
Redundancy problem 155-175, 198
Relaxation 60-65, 82-83, 108-110, 217-223, 230, 232, 235-236, 239-240, 242, 256
overdamped 82
time 60-65, 108-110

S

Scale free 202
Schemas 84
Segregation 198-199, 211-215, 243-259
Selection 177, 193, 249-250

Self-organization 21-22, 36-37, 54, 83, 86, 103, 105, 107, 110, 113, 116-117, 177-178, 193, 233, 240, 258
Semantics 7-9, 16
Sensorimotor 19, 141-142, 144, 149, 198, 233, 248

Sensory
  coordinates 3
  language 3
  neurons 227, 284
Simulator 15, 17
Small world 202, 212, 214-215
Spatial array 2-3, 11
Spatial temporal grouping 253
Specification 28, 156-160
Spiking 219-220, 235
Stability
  behavioral 31
  bistable 64, 67-68, 71, 73, 80-81, 246, 250, 257
  metastability 197-199, 201-213, 235
  monostable 23, 246, 250
  multistable 14, 68, 73, 82
Stabilization 84, 91-101, 114-115, 120, 138, 257
Stochastic 23, 38, 60-64, 86-87, 99, 126-131, 135, 138, 201, 220, 228-229, 241
  process 126-127, 138
Strategy
  neural 109
  postural 105, 158
Streaming 245-251
Stretch reflex 115, 125, 158, 160-161, 174
Symbols 3, 15, 16, 58
Symmetry 66, 68-69, 71, 73, 83-84, 179-192, 249, 251, 256
  asymmetry 184
  breaking 23, 250-251
  conservation 58-59
Synchronization 38, 143, 153, 180, 183, 189, 206, 214-215, 231, 234-235, 240-241, 248-250
Synchronous firing 227, 239
Synchrony 198, 206, 215, 118-119, 222, 227, 233-236, 238-239
Syncopation 36, 38, 144, 248, 249
Synergetics 13, 19, 85, 257
Synergy 147, 163, 168-169

System nervous 19, 21, 22, 99, 109, 125, 136, 141, 148, 156-157, 160, 162-164, 173-175, 213, 215
Systems
  afferent 157
  biological 21, 59, 116, 158, 175, 178, 254
  controllable 104
  dynamical 14, 16, 21, 36, 74, 83, 104, 119, 124, 158, 193, 202-203
  formal 17
  natural 17
  musculoskeletal 125, 142-145, 147-148, 152, 162-163, 170, 178, 192-193, 256
  non-autonomous 12
  non-equilibrium 59, 101, 110
  perception-action 8, 14
  perceptual-cognitive 177-181, 183-192, 195, 258
  postural 103-105, 107, 110-111, 113, 116, 127

T
Tasks
  cognitive 198, 233
  demands 48, 58, 65, 119, 171, 175
  dual 25-27, 29-30, 36, 44, 46, 79-80, 88, 145
  motor 25, 58, 151, 171
  supra-postural 104, 110
Temporal judgments 232
Temporal processes 244
Thalamo-cortical circuits 211, 235-237, 239, 241
Theory 5, 8, 14-16, 19, 20, 38-39, 42, 57-59, 87-89, 91, 101, 104, 109, 110, 113, 120, 124, 136, 139, 175, 193-194, 212, 215, 223-224, 229, 232, 239-240, 242, 245, 259
  control 91, 124, 136
  dynamic pattern 42, 87

Time scales 11, 21, 58-59, 66, 77, 79, 82, 84, 87, 93, 97, 99, 114, 123, 134, 221-22, 230, 234
Time-delay 129-130, 251
Time-to-balance 92-93, 95, 97, 99-101
Time-to-contact 92
Timing 38, 43, 46, 51, 150, 218, 232, 245, 247
Trajectory 43, 61 82, 117, 124-130, 137, 150, 155, 163, 168-175, 235, 243, 250, 252-255
   convergence 117, 243, 253-255
   formation 117, 243
Transfer 39, 50-51, 54, 68-71, 84, 86, 89, 119, 121, 124, 226, 235
Transient 79, 81, 107, 164, 201, 204, 211, 214, 235, 243, 249-251
Transient adaptability 79, 81
Transition
   phase 31-34, 37-39, 49, 54, 56, 60, 80, 87, 89, 101, 103, 114, 117-118, 120-121, 152-153, 183, 193-195, 249, 256-258
   vicinity 3, 108, 249
Translation 3, 98, 117, 195, 259

U
Unit 3, 5, 6, 8, 14, 62, 103, 11, 133, 142-144, 159, 173, 175, 200, 202-205, 210-211, 223, 255
   motor 142-144, 159, 173
   single 14, 103

V
van der Pol model 218-219, 223, 230, 242
Variability 22, 26-34, 58, 68, 74-82, 107, 110, 115, 125
Variables
   collective 13, 14, 22, 28, 62, 104
   control parameter 23, 42, 108, 156, 178, 244-248
   dependent 235
   order parameter 13, 118, 193, 249
   perceptual 95, 97, 98
Vision 1, 14, 18, 55, 110, 113, 117, 129, 130, 138-139, 149-151, 157, 173, 183, 244
Volition 21, 48, 53